Orthogenesis versus Darwinism

Igor Popov

Orthogenesis versus Darwinism

Translated by
Natalia Lentsman
Saint Petersburg, Russia

Springer

Igor Popov
Saint Petersburg State University, N. N. Petrov Research Institute of Oncology
Saint Petersburg, Russia

ISBN 978-3-030-06983-4 ISBN 978-3-319-95144-7 (eBook)
https://doi.org/10.1007/978-3-319-95144-7

This Springer imprint is published by the registered company Springer Nature Switzerland AG
The registered company address is: Gewerbestrasse 11, 6330 Cham, Switzerland

Contents

1 Introduction 1

Part I Orthogenesis: A History

2 The Sources. Formation of the Concept of Directed Evolution in the Nineteenth Century 7

3 Evolutionary Biology at the Turn of the Twentieth Century. New Concepts of Directed Evolution in the 1900–1930s 31

4 Declarations in Favour of Orthogenesis in the 1900–1930s 53

5 Orthogenesis and the Modern Evolutionary Synthesis 73

6 New Concepts of Directed Evolution in the Recent History of Evolutionary Biology 85

7 Statements in Support of Orthogenesis in the 1950s and Later 101

Part II Orthogenesis: Pro et Contra

8 "Laws" in Biology 117

9 Species Senescence 129

10 Adaptation or Non-adaptation? 139

11 Anticipation, Parallelisms and Convergences 151

12 Constraints on Variation 159

13 Direct Impact of the Environment on Evolution and the Inheritance of Acquired Characteristics 177

14 Mystics or Unrealised Synthesis? 187

15 Conclusion 201

Index 207

Chapter 1
Introduction

> *Occupying oneself too much with history may make one unproductive but ignoring history means saying much of what has already been said, and probably much better said, before.*
>
> *Hans Driesch. Preface to the first edition of Der Vitalismus als Geschichte und als Lehre (The History and Theory of Vitalism), 1905. (Here and elsewhere in this book, citations from works in languages other than English were translated by Natalia Lentsman, unless stated otherwise.)*

Darwinism is based on the idea that evolution occurs due to selection from a large or an almost unlimited source of variability and that its direction is determined by adaptation to a constantly changing environment. An alternative viewpoint implies that the material of variability is limited, that organisms are predisposed to vary in certain directions and that this predisposition is the determinant of evolution. While evolution does respond to external changes, it does not depend on them; adaptation is not the evolutionary mainstream. This viewpoint has usually been referred to as *orthogenesis* (from the Greek *orthos*—straight), that is, development in a certain direction or "directed evolution". Numerous other terms have also been coined to express this idea (nomogenesis, autogenesis, ologenesis, autoevolution, to name a few), while the term "orthogenesis" has not always meant directed evolution. Not uncommonly, it was used to describe a phenomenon when related groups of organisms (in fossil record or in current existence) could be said to represent series of directed transformations. Orthogenesis is closely associated with the doctrine of organicism, which postulates that species or higher taxonomic groups undergo the same life cycle as an individual: birth, growth, development and natural death. Though Darwinism holds the ideas of orthogenesis and organicism to be erroneous, their followers have accumulated some thought-provoking facts, which it would be unreasonable to ignore.

The ideas of directed evolution have a bad reputation. In the popular opinion, they smack of vitalism, mystics, teleology, idealism and other similarly ignoble concepts. This is all the more surprising as the advocates of orthogenesis usually

I. Popov, *Orthogenesis versus Darwinism*,
https://doi.org/10.1007/978-3-319-95144-7_1

went to another extreme, tending to reduce biological phenomena to the processes studied by physics, chemistry and other natural sciences. They also used to search for "periodical systems" of living organisms and various biological "laws". A graphic example of this approach is the comparison of evolution with the growth of crystals. According to the concept of orthogenesis, evolutionary trends follow some internal laws just as crystals grow to acquire a certain shape.

The ideas of directed evolution date back to antiquity but orthogenesis as an articulated evolutionary concept emerged in the second half of the nineteenth century and was actively developed in the first three decades of the twentieth century. With the advance of modern Darwinism in the middle of the twentieth century, orthogenesis dwindled, overshadowed by the natural selection theory. It has not disappeared, however, and continues to attract new followers.

This study offers an analysis of orthogenesis in historical perspective. As far as it is possible in a historical study, I have attempted to assess the arguments offered by the supporters of orthogenesis and its critics. The key question here is whether orthogenesis holds something valuable for modern biology or is, as many Darwinians have claimed, a rotten idea and a dead end of the evolutionary thought.

There are fewer historical studies of orthogenesis than of other evolutionary concepts. Ideas of directed evolution are usually mentioned insofar as they are relevant to works of particular scientists, certain historical episodes or the history of evolutionary biology in general. In Russia, the first review paying considerable attention to directed evolution was published by Yuri A. Filipchenko in 1923 (Filipchenko 1977). A remarkable feature of this work was that the author tried to evaluate facts and concepts impartially, without trying to assert his own views about evolutionary mechanisms. He expressed them, of course, but did not focus on them. This work, though naturally out of date, still reads surprisingly well.

As Darwinism came to dominate evolutionary biology, the prevailing mood of its followers was to "defend" it from various "assaults". Orthogenesis, being one of them, was hardly ever interpreted objectively. This was especially typical of studies conducted in the USSR, where Darwinism was supported by the state ideology. Orthogenesis was usually mentioned in passing but some works contained a rather extensive material on it (Davitashvili 1948, 1966; Zavadsky 1973). With the weakening of ideological pressure, a less critical approach to non-Darwinian concepts became possible. Discussions of directed evolution can be found in the studies of Vadim I. Nazarov devoted to the evolutionary theory in France, finalism and macroevolution (Nazarov 1974, 1984, 1991). Some aspects of directed evolution were analysed by Daniil A. Alexandrov in his research of the history of morphology in Russia (Alexandrov 1996).

Western historians and biologists tended to ignore rather than criticise the ideas of directed evolution. Most historic studies dealing with orthogenesis were authored by Peter Bowler (Bowler 1979, 1983, 1989). They were devoted to non-Darwinian theories of the early twentieth century, the American school of evolutionists of the second half of the nineteenth century, the works of Theodor Eimer, and the analysis of the evolutionary idea as a whole. Bowler remarked that, at the time when he had started his studies, many of his colleagues had considered non-Darwinian concepts as “side-lines” of “peripheral importance” or even “odd theories” of “merely antiquarian interest” (Bowler 2003, p. 1). To note, Bowler’s studies were restricted to the literature published in English (the works of Eimer being available in the English translation).

Robert Richardson and Thomas Kane addressed the formation of the orthogenesis concept in the nineteenth century (Richardson and Kane 1989). Their study, however, was incomplete and relied solely on American sources. Steven Jay Gould paid orthogenesis relatively much attention in his last book (Gould 2002) but characterised this concept based on a rather arbitrary selection of the authors (Th. Eimer, A. Hyatt, Ch. Whitman). Wolf-Ernst Reif discussed directed evolution in the context of the history of German palaeontology and some other disciplines (Reif 1983, 1986, 1999). Some aspects of the history of orthogenesis were analysed by O. Rieppel (2012) and G. Levit and L. Olsson (2006).

These publications and several general reviews on the history of biology were used as an initial point of the present study (Mikulinsky 1972; Blyakher 1975; Mayr 1982; Jahn 2000). Based on them, I compiled the bibliography of primary sources on orthogenesis. These sources were studied with a focus on the biological content, with the analysis of biographies and other historic circumstances being reduced to a minimum. The choice of such an old-fashioned methodology was associated with the main aim of this study: to pick up from the wealth of historical data ideas relevant in the context of modern biology. The material for the study was mostly collected in the libraries of St Petersburg (Russia), Paris (France) and Vienna (Austria).

Working on this book, I found it reasonable to present separately the history of the concepts of directed evolution and the arguments of their advocates. This is due to the fact that the essence of the polemics between Darwinism and orthogenesis has hardly changed over its history, with the main arguments being repeated over decades in almost identical form. A detailed analysis of argumentation of each author in chronological order would result in numerous repetitions. Nevertheless, I failed to avoid repetitions altogether because the history of orthogenesis turned out to be exceptionally entangled.

Acknowledgements I am grateful to the Konrad Lorenz Institute for Evolution and Cognition Research and Fondation Maison des science de l’homme for providing financial support of the studies (in 2005–2007 and in 2013) that resulted in this book. I am grateful to Natalia Lentsman and Springer Publishing Company for careful work with my manuscript. I thank Dr. Andrew Ostrovsky, Saint Petersburg State University, for advice and encouragement during preparation of the manuscript.

References

Alexandrov DA (1996) The history of Russia evolutionary moprhology in the end of the 19th – the first third of the 20th century (Istoriya otechestvennoy evolyutsionnoy morfologii zhivotnykh v kontse XIX v. – pervoy treti XX v.). Extended abstract of Cand Sci (Biol) Dissertation. Moscow

Blyakher LYa (ed) (1975) History of biology from the beginning of the 20th century to the present (Istoriya biologii s nachala XX veka do nashikh dney). Nauka, Moscow
Bowler P (1979) Theodor Eimer and orthogenesis. J Hist Med 34:40–78
Bowler PJ (1983) The eclipse of Darwinism. Johns Hopkins Univ Press, Baltimore
Bowler PJ (1989) Evolution. The history of an idea. Univ of California Press, Berkeley
Bowler PJ (2003) Introduction (Predislovie). In: in the shadow of Darwinism. Alternative theories of evolution in the 20th century (V teni darvinizma. Al'ternativnye teorii evolyutsii v XX veke). Yasnyy den', St Petersburg, p 5–6
Davitashvili LSh (1948) The history of evolutionar palaeontology from Darwin to our days (Istoriya evolyutsionnoy paleontologii ot Darvina do nashikh dney). AN SSSR, Moscow, Leningrad
Davitashvili LSh (1966) The modern state of evolutionary teaching in the west (Sovremennoe sostoyanie evolyutsionnogo ucheniya na Zapade). Nauka, Moscow
Driesch H (1905) Vitalism as history and as theory (Der Vitalismus als Geschichte und als Lehre). Verlag von Johann Ambrosius Barth, Leipzig. English edition: Driesch H (1914) the history and theory of vitalism (trans: Ogden CK). Macmillan, London
Filipchenko YuA (1977) Evolutionary idea in biology: a historical review of evolutionary teachings of the 19th century (Evolyutsionnaya ideya v biologii: istoricheskiy obzor evolyutsionnykh ucheniy XIX veka). Moscow
Gould SJ (2002) The structure of evolutionary theory. Harvard Univ Press, Cambridge
Jahn I (ed) (2000) The history of biology. Theories, methods, institutes, concise biographies (Geschichte der Biologie. Theorien, Methoden, Institutionen, Kurzbiographien). Spektrum, Heidelberg/Berlin
Levit GS, Olsson L (2006) Evolution on rails: mechanisms and levels of orthogenesis. Ann Hist Phil Biol 11:97–136
Mayr E (1982) The growth of biological thought. Harvard Univ Press, Cambridge (Mass)
Mikulinsky SP (ed) (1972) The history of biology from ancient times to the early 20th century (Istoriya biologii s drevneyshikh vremen do nachala XX veka). Nauka, Moscow
Nazarov VI (1974) Evolutionary theory in France after Darwin (Evolyutsionnaya teoriya vo Frantsii posle Darvina). Nauka, Moscow
Nazarov VI (1984) Finalism in the modern evolutionary teaching (Finalizm v sovremennom evolyutsionnom uchenii). Nauka, Moscow
Nazarov VI (1991) The teaching about macroevolutions. On the ways towards a new synthesis (Uchenie o makroevolyutsii. Na putyakh k novomu sintezu). Nauka, Moscow
Reif W-E (1983) Evolutionary theory in German paleontology. In: Grene M (ed) Dimensions of Darwinism. Cambridge Univ Press, Cambridge, p 173–203
Reif W-E (1986) The search for a macroevolutionary theory in German paleontology. J Hist Biol 19:79–130
Reif W-E (1999) Deutschsprachige Paläontologie im Spannungsfeld zwischen Makroevolution und Neodarwinismus (1920–1950). Verhanglungen zur Geschichte und. Theorie der Biologie 2:151–188
Richardson RC, Kane TC (1989) Orthogenesis and evolution in the nineteenth century. In: Nitecki M (ed) Evolutionary progress. Univ of Chicago press, Chicago/London, pp 149–169
Rieppel O (2012) Karl Beurlen (1901–1985), nature mysticism, and aryan paleontology. J Hist Biol 45:253–299
Zavadsky KM (1973) The development of evolutionary theory after Darwin (1859–1920s) (Razvitie evolyutsionnoy teorii posle Darvina [1859–1920-e gody]). Nauka, Leningrad

Part I
Orthogenesis: A History

Chapter 2
The Sources. Formation of the Concept of Directed Evolution in the Nineteenth Century

Abstract The pre-history of the concepts of directed evolution is represented by treatises of the authors who worked in the first half of the nineteenth century (Jean-Baptiste Lamarck, Georges Cuvier, Giovanni Batista Brocchi and others). They contained ideas related to spontaneous evolution, constraints on variation and the similarity of lifecycles of species and those of individuals. In the second half of the nineteenth century the concept of directed evolution was developed by Albert von Kölliker, Carl Nägeli, Edward Cope, Alpheus Hyatt, Wilhelm Haacke and Theodor Eimer, who worked more or less independently of each other. Each of these authors can be credited with being, in some respect, the pioneer of orthogenesis. The first works dealing with directed evolution were disconnected, and we cannot find in them the first wording of this concept, however hard we might try. This elusive concept can be deduced from the context of the writings of several authors, who claimed that organisms changed spontaneously in definite directions and denied the leading role of natural selection. As for the rest, their ideas had little in common.

Keywords Lamarckism · Purposeful evolution · Non-adaptive characters · Constraints on variation · Nineteenth century

The history of evolutionary biology dates back to the remote past. One can find evolutionary ideas in the poetry of Lucretius, the treatises of Aristotle and Empedocles, even in ancient myths. However, when it comes to orthogenesis, it seems reasonable, following the advice of Hans Driesch (see Epigraph), not to occupy oneself too much with history of science and associated fields of knowledge. The idea of directed evolution as understood in the present study implies that the knowledge about the living world has reached a stage when evolution is an object of scientific research.

In this case, the starting point of the historic analysis lies in the treatises dating back to the early nineteenth century, first and foremost, those of **Jean-Baptiste Lamarck** (1744–1829). Indeed, orthogenesis has become so closely associated with the ideas of Lamarck that it is sometimes considered as a component of Lamarckism or a version thereof. There have even been attempts to reveal a "true" orthogenesis—that of Lamarck himself (Steinmann 1908). This "true" orthogenesis

I. Popov, *Orthogenesis versus Darwinism*,
https://doi.org/10.1007/978-3-319-95144-7_2

has, however, little to do with the concept characterised in the introduction. It belongs to the prehistory of ideas concerning directed evolution. Though Lamarck did express an idea about spontaneous evolution irrespective of environmental changes, he mainly analysed other evolutionary mechanisms. One would look in vain for any exact wording of a concept close to orthogenesis in his treatises. The two major evolutionary laws suggested by Lamarck concern exercises and inheritance of acquired characters. Most of his famous *Philosophie Zoologique* (Philosophy of Zoology) also deals with other subjects.

Some allusions to ideas reminiscent of orthogenesis can be gleaned from meticulous scrutiny of certain pages of Lamarck's writings. A chapter of *Philosophie Zoologique* dealing with degradation seems to be consistent with orthogenesis. Lamarck called attention to the impressive series of decreasing or increasing structural complexity of all systems of organs (gradation or degradation) that could be observed in animals. Yet even in this chapter the idea of spontaneous transformation irrespective of the changes of the environment is implied rather than stated: "It is obvious that, if nature had given existence to none but aquatic animals and if all these animals had always lived in the same climate, the same kind of water, the same depth, etc., etc., we should then no doubt have found a regular and even continuous gradation in the organisation of these animals" (Lamarck 1809 cited in Lamarck 1914, p. 69. English translation by Hugh Samuel Roger Elliott). This means that, according to Lamarck, evolution can be driven by internal processes only, without the impact of environmental changes. The mechanism of this transformation is "nature", "general and particular laws governing this order" etc. This is as close as Lamarck gets to putting into words the idea of directed evolution, and these vague speculations pass almost unnoticed in this treatise.

A striking difference between Lamarck's idea about gradation or degradation and most other evolutionary concepts (including those concerning directed evolution) is that Lamarck did not represent evolutionary ways as branching genealogical trees. Lamarck (and his predecessors) described evolution as straight lines originating (from non-living world) spontaneously and independently at different times. These different times of origin would explain the varying complexity of simultaneously occurring organisms. It should be noted, however, that a simple branching phylogenetic scheme with two initial points was presented elsewhere in the same treatise (in the appendices to *Philosophie Zoologique*). By that time, the possibility of representing the variety of organisms in form of trees had already been suggested. **Peter Simon Pallas** (1741–1811) discussed various kinds of representing the system of organisms in his work *Elenchus Zoophytorum* (1766) and came to the conclusion that a tree was the best one (cited in Raykov 1952). Apparently, Lamarck was unfamiliar with the works of Pallas. He thought about genealogical schemes independently, and his studies in this direction were left unfinished.

Lamarck was much more interested in various deviations from gradation occurring because of an unlimited variety of the organisms' living conditions. He paid them most attention in *Philosophie Zoologique*, carefully analysing their mechanisms such as "circumstances", "exercises", "habits" and "use and disuse". Reading expressive circumlocutions about profound changes caused by these processes, one forgets entirely that these are mere deviations from gradation. Lamarck made these deviations seem at least as important in evolution as gradation itself, if not more so. It is for this reason that the concept of exercises, habits and direct action of the environment upon evolution has traditionally been considered as the most important aspect of Lamarckism.

After *Philosophie Zoologique* Lamarck probably continued to think about spontaneous transformation. With time he changed the wording of his laws and proposed four laws instead of two. One of them suggests orthogenesis: "Life by its proper forces tends continually to increase the volume of every body possessing it, and to enlarge its parts, up to a limit which it brings about" (Lamarck 1822, translation into English: http://www.nndb.com/people/275/000057104/). This claim, however, is blurred and, being surrounded by nearly opposite claims, cannot be considered as a foundation of an evolutionary theory. At the same time, Lamarck's writings lack many of the ideas usually associated with orthogenesis such as criticism of adaptation and positing constraints on the organisms' ability to change. This indicates that Lamarck was the author of a single element of the orthogenetic concept, namely, the idea of spontaneous transformation, and even this idea was expressed in the most general terms.

The works **George Cuvier** (1769–1832), Lamarck's major antagonist, also contain indications of a nascent orthogenesis. Some of them are even more definite than those that can be found in Lamarck's writings, and several historians are quite sure that Cuvier laid down the basis for orthogenesis (Devillers and Chaline 1993). The authors of the Frankfurt construction theory, a recent concept of directed evolution, also consider Cuvier as a predecessor while ignoring most other possible candidates (Gutmann 1996). This is due to the fact that orthogenesis is based on idea about constraints of variation, and Cuvier was one of the first to address this phenomenon. Cuvier is known for his achievements in the reconstruction of extinct animals and other osteological studies. A popular legend has it that a tooth was enough to him to reconstruct the shape of its owner. This kind of research is certainly based on the idea that all structures of an organism are strongly correlated. According "to the law of correlation of the forms", the number of combinations of characters is limited, while some combinations are "forbidden" (for example, there are no carnivorous ungulates and no carnivorous animals with horns). Variability is limited, and there is predisposition to certain variants. This is exactly what the advocates of orthogenesis claim.

It is known that Cuvier did not discuss his materials in evolutionary terms though he came close to this interpretation. He wrote about the change of faunas in the history of the Earth noting that as some animals became extinct in an area (probably, as a result of catastrophes), some other animals appeared there (Cuvier 1850). How exactly the new ones originated remained an open question. Cuvier mentioned in passing that they might arrive from other areas. Though Cuvier was certainly no evolutionist, he is sometimes considered a harbinger of evolutionary biology, because he introduced order into studies of fossil animals, establishing a solid bases for evolutionary analysis (Devillers and Chaline 1993; Kanaev 2000). His paleontological works were probably more important for the advent of evolutionary biology and, in particular, the concept of directed evolution than speculations about spontaneous transformations.

There was at least one other contemporary of Lamarck and Cuvier who was also a precursor of orthogenesis. An Italian palaeontologist **Giovanni Batista Brocchi** (1772–1823) suggested that species or other taxa underwent the same life cycle as individuals: birth, growth, heyday, decline, decay and death. He wrote eloquently that, as a group of organisms evolves, its vitality decreases, development weakens, growth stalls, fertility and ability to reproduce dwindle etc.; finally, it comes to the fatal end (Brocchi 1843, p. 412). This idea is very close to orthogenesis as it asserts internal mechanisms of evolutionary development, while contradicting the basic postulates of Lamarckism, which denied extinction and dramatic evolutionary changes.

This indicates that some components of orthogenesis were already present in the first half of the nineteenth century but remained almost unconnected. The next step was made in the late 1860s, when the theory of natural selection gave a powerful impetus to biology. Several authors presented the idea of internally directed evolution almost simultaneously: as an alternative to Darwinism, as an addition to it or just as a summary of their empirical research seemingly unconnected with Darwin's ideas.

Carl Wilhelm von Nägeli (1817–1891), a Swiss-German botanist, was one of the pioneers of directed evolution. His research interests were cell division, the structure of cell wall, the taxonomy of the family Hieracea, fermentation and the biology of microorganisms (Schmidt 2000). As early as in 1856, that is, before the publication of *The Origin of Species*, Nägeli published a paper on individuality expressing some ideas in terms of directed evolution (Nägeli 1856). He considered individuality, i.e. an entity that could not be decomposed without destroying its existence, as a general characteristic of natural phenomena such as spiritual life, living organisms and chemical substances. Arguing against teleology, he tried to find proof that living and non-living matter was subordinate to the same laws. A popular argument of that time against this viewpoint was that cells and tissues could be created from inorganic substances but Nägeli did not consider it as valid. He thought that humans could not create anything in inorganic world either; they could just allow some processes to be manifested. Nägeli enthusiastically supported the idea of evolution but did not analyse it profoundly. ("Species exist for a period of time, then they are replaced by others. We can only guess how they originate. There is a viewpoint that previously existing species become extinct and new ones originate, either in form of embryo or even fully formed. Like Pallas Athena came from the Zeus' head, so an oak with its branches and leaves or an elephant with its tusks and trunk emerged in the nature. According to the second viewpoint, a new species originates from a previously existing one. Theory and practice testify in favour of the second viewpoint." (Nägeli 1856, p. 203). Later, with the advent of the idea of evolution in biology, Nägeli participated in discussions concerning the theory of natural selection. In a paper published in 1865 he offered his new concept considering it as an addition to Darwin's theory.

In some respect Nägeli was close to Lamarck. He also considered evolution as a spontaneous process developing in the progressive direction. Following Lamarck, Nägeli explained simultaneous existence of organisms at various stages of evolution by the multiple origin of life. He decided to supplement the "principle of utility" represented by adaptations and argued against the view that variation could proceed in all directions and that evolution was determined only by the impact of the environment. Nägeli mostly relied on proof by contradiction and offered the following arguments:

1. According to the "theory of usefulness" one could expect a single form to originate in a certain environment, yet in reality many various forms can exist under the same conditions.
2. According to the "theory of usefulness", the transformation of a species should cease once it has reached an adapted condition. In fact, a species continues to evolve even in an unchanging environment, its evolution being driven by an "urge for perfection".
3. If an organism adapted to a certain environment finds itself in a new one, it changes and becomes adapted to new conditions. This means that if it returns to the initial environment it should reverse to the initial state. However, it is well known that feral forms do not become similar to their wild ancestors.
4. One may expect neutral characters to be variable and useful ones to be stable. In reality, an opposite situation is observed. Characters underlying the separation of plant groups of high taxonomic level (e.g. phyllotaxis) are neutral but stable.

Observations like these prompted a conclusion that "the theory of perfection should be recognised in addition to the theory of utility" (Nägeli 1865). The "theory of perfection" was based on the assumption that evolutionary changes do not proceed in all directions. They follow only certain routes, mainly progressive ones, and these depend not only on external causes.

Another Swiss-German scientist, **Rudolf Albert von Kölliker** (1817–1905), suggested similar ideas almost simultaneously with Nägeli. An embryologist and a histologist, Kölliker was known for his research on sperm, eggs of cephalopods and taxonomy of pennatulids (Schmidt 2000). In his prolific studies Kölliker was guided by the idea that evolution, as well as everything else studied by biologists, was based on "the laws of organic nature" and reproached Darwin and his followers, especially Ernst Haeckel, for ignoring their existence. In the 1860s Kölliker presented a concept of "heterogenesis".[1] Its main difference from Nägeli's theory was that Kölliker not only postulated the existence of directed evolution but also developed ideas about drastic evolutionary changes. For this reason Kölliker's concept is often considered as one of the first versions of saltationism or "evolution by leaps".

[1] The word "heterogenesis" has later been used to name at least two other biological theories: an evolutionary theory close to mutationism-saltationism (Korzhinsky 1899) and a theory unrelated with evolution and concerning the relations of phylogenesis and ontogenesis (Kryzhanovsky 1939).

Kölliker analysed this phenomenon quite profoundly. His heterogenesis is close to the idea of "ontomutations" suggested much later by Albert Dalcq, that is, of mutations affecting early stages of embryogenesis and resulting in the occurrence of new groups of high taxonomic level (Dalcq 1954). Kölliker focused on the events when "following general laws of nature, animals developed from their eggs new creatures different from themselves" (Kölliker 1864, p. 942). The heterogenesis can be illustrated by instances of "alternating life forms" in the life cycle of some invertebrates: jellyfish transforming into polyps, echinoderms giving birth to free-swimming larvae. In these instances, higher forms seem to produce lower ones. According to Kölliker, this process shows how higher forms evolved. This means that he shared the idea of recapitulation. Kölliker also expressed an idea about the evolutionary role of neoteny (reproduction at larval stage).

Kölliker noted that evolution by leaps was not the most important component of his concept. It concerned only one aspect of evolution. He believed that gradual evolution was also possible, taking into account that any evolutionary transformation was based on "the laws of development". These laws, he thought, were the antithesis of the teleological idea of usefulness. Kölliker claimed that Darwin was a true teleologist since he considered any minute detail of an organism as created for its benefit. Kölliker hold such non-materialistic views to be unacceptable. He was also displeased by some other aspects of Darwin's theory. Kölliker pointed to instances of a prolonged existence of some invertebrate species, which did not seem to evolve at all for a long time, and to gaps in fossil record. Besides, he believed that nobody had ever observed the transition from a variety to a new species. In general, Kölliker considered the explanation of evolution by natural selection as absolutely unfounded. In his view, it was similar to explaining the origin of new chemical elements by selection of minute deviations. He indicated that "series of forms" in fossil records could be considered as analogical to series of chemical substances suggested by chemists. However, despite all his criticism of Darwin's theory, Kölliker still claimed that his own theory was similar to it. Heterogenesis just eliminated the notions of "usefulness" and "the fittest" varieties.

Later Kölliker continued to develop his ideas on evolution. He renamed his "heterogenesis" into the "theory of development under the influence of internal causes" (Kölliker 1872), emphasising the significance of the "laws of development". According to Kölliker, the general laws of organic development were not different from what was meant by "a researcher of minerals speaking about the laws of growth of crystals or by an astronomer speaking about the laws of gravitation or the laws of development of heavenly bodies" (Kölliker 1872, p. 4). His work (Kölliker 1872) abounds with pronouncements of this kind, repeated, with slight variations, with the insistency of a conjurer casting spells. Finally, he concluded that it was unnecessary to dwell further on "the laws" because many great scientists such as Nägeli, Heer, Rütimeyer, His and others had already pronounced their opinions on this topic. We can judge from this evidence that such views were popular among biologists of that time. Kölliker considered the existence of these laws as evident and was bewildered by the fact that they did not attract as much attention in biology as in other natural sciences.

A short time after Kölliker and Nägeli, ideas of directed evolution were suggested by American palaeontologists **Edward Drinker Cope** (1840–1897) and **Alpheus Hyatt** (1838–1902). Their pronouncements on evolution triggered the development of the American Neo-Lamarckism school of thought. This school formed in isolation from European evolutionists and their immediate predecessors. Its similarity to Lamarck is likely to be convergent since Neo-Lamarckians hardly ever cited Lamarck; in fact, it is probable that they had not even read his treatises (Pfeifer 1965). In many respects American Neo-Lamarckians were much closer to Lamarck's antagonist, Georges Cuvier. Their teacher, **Louis Jean Rodolphe Agassiz** (1807–1873), had been a student of Cuvier and developed similar research topics in America (Richardson and Kane 1989). Agassiz was even further from evolutionary theory than Cuvier but he was an outstanding palaeontologist. His American students used the experience of the Cuvier school in their evolutionary studies.

The first representative of the "American school" to publish a paper on directed evolution was Alpheus Hyatt. He studied extinct organisms, mainly molluscs, and similarly to J. Brochhi came to the conclusion that the life cycle of a species reminded that of an individual. Hyatt also suggested the idea of recapitulation and evolution by saltation (Hyatt 1866). Very soon after that, Edward Drinker Cope expressed his full support of Hyatt and suggested several other ideas (Cope 1868). Cope, who worked with fossils, stated several laws characterising paleontological material. The most important of them were as follows. According to the "law of acceleration", a character of an organism emerges at an earlier stage of its ontogenesis than in its ancestor. (Cope considered this as a "principle" or a mechanism of progressive evolution: since organisms acquire more and more new characters in the course of evolution, ancestral characters should emerge earlier to allow a successful completion of the ontogenesis). The "law of retardation" makes provision for regressive evolution: the emergence of a character at the later stage of an organism's ontogenesis than in its ancestor. In other words, in case of a regressive evolution an organism acquires "progressive" characters and then reverts, acquiring the final characters later than the ancestor. Later Cope specified that retardation was not always exact and so it was incorrect to refer to the process of regressive evolution as to retardation. He suggested the term "senescence" for this case (Cope 1896).

Speculating about retardation and acceleration, Cope also proffered the principle of the non-specialised ancestor (often referred to as the "Cope's rule"). According to this rule, in the beginning of a group's evolution there are non-specialised organisms, which specialise as they evolve. Finally their specialisation becomes so narrow that the group can no longer adapt to environmental changes fast enough and becomes extinct. Cope considered his "laws" as an indication of the directed evolution contrasting with Darwin's theory.

In 1868 Cope published a rather peculiar paper entitled *On the Origin of Genera* (Cope 1868). Its very title implies that the explanation of the origin of species is not enough to build an evolutionary theory. Species are parts of higher systems and cannot be arbitrary. Cope believed that the same species could originate in different genera and tried to reveal the rules of their origin. Much of this work deals with the analysis of rules governing variation of animals. Cope put much effort in attempts to provide a theoretical background for the taxonomy of living organisms. He even tried to use the data from chemistry comparing the variety of organisms with that of chemical elements and substances. He likened series of fossil forms with series of chemical elements arranged in the order of increasing or decreasing oxidizing properties. The periodic system of elements was yet unknown but Cope apparently had something similar in mind. He wrote about the classification of genera: "It is precisely as, among the inorganic elements, we pass from the electro-negative, non-oxidizing extreme of the Halogens, with Fluorine as the extreme, to the electro-positive, violently oxidizing extreme of the alkaline metals, whose extreme is potassium, by steps whose relative position is measured or determined first by these tests; and as these steps have each their included series of bodies, characterized by their successive relations on the lower level of a subordinate range of characters." (Cope 1868, p. 244). This means that oxidising or deoxidising capacities are the major criteria for the classification of chemical elements, which can be used to identify groups and subgroups. These groups and subgroups were, according to Cope, analogous to genera and species. A kind of "chemical classification" of organisms could be created.

At that time Cope mainly studied extinct vertebrates. Unfortunately, no general criteria similar to oxidising properties of elements could be revealed in this material. As a result, Cope's systems, with their description of parallelisms (e.g. in the evolution of horns or hoofs), resemble fragments of chemical classifications. Cope tried to establish the limits of parallelisms, wondering whether all characters of organisms could be described by parallel series. Based on the data at his disposal, Cope concluded that it was not so. Parallelisms could be complete or incomplete.

In addition to the analysis of parallelisms, Cope composed numerous tables with combinations of various characters of animals (features of the circulatory system, the skeleton etc.). In these tables he marked existing and non-existing combinations. To some extent, these tables anticipated homologous series in biology or the periodic system in chemistry. In practice, however, they turned out to be cumbersome and inconvenient, and were ever hardly used by Cope or his followers.

Alongside these typological forays, Cope did some serious analysis of evolution. He disagreed with virtually every aspect of Darwin's natural selection theory. He looked for the mechanisms of evolution in some internal laws of development unrelated to adaptation and selection. Cope believed that adaptation was an insignificant aspect of evolution. Characters underlying differences between genera originated otherwise, under the impact of some internal forces. Selection cannot cause primary changes, it can merely support them should they prove useful. Primary changes are driven by "growth energy", "necessity", "desire" etc.

Afterwards Cope wrote several other evolutionary studies analysing some other mechanisms in addition to internal forces such as bathmism. In particular, he shared Lamarck's ideas about exercises and habits (Bowler 1983, 1989; Richardson and Kane 1989). This seems strange because exercises and internally predetermined evolution had always been in opposition (see the above discussion about Kölliker and Nägeli) but Cope, apparently, saw no contradiction there.

An analysis of some passages from his work clarifies this situation. Discussing evolution, Cope and his colleagues frequently resorted to the terms "definite variation" or "direct variation". In the biological context, these terms are ambiguous: they may refer to the variation determined by the environment or to the predetermined variation. Darwin wrote about definite variability meaning that it was not inherited but determined by the environment. At the same time, "definite" may mean predetermined, internally directed, "orthogenetic" variability determined by some obscure structural or developmental laws rather than the environment. It is quite impossible to clarify this matter because these terms could be used in different sense even within one and the same paper (Definite vs. Fortuitous variation... 1892). This ambiguity is likely to have arisen because Cope and his colleagues (Alpheus Spring Packard, 1839–1905; Joel Asaph Allen, 1838–1921) were strongly opposed to the idea of random variability (what we now call mutations). They were united in the opinion that organisms varied not randomly but in certain directions, and promulgated it in every possible way. Whether these directions were determined by internal or external causes was a matter of lesser importance for them though they did tend to accept Lamarckian ideas about exercises and nervous pulses. Some of the American Neo-Lamarckians even expressed extremely ectogenetic ideas. Joel Allen, who conducted an extensive research on geographic variability of mammals and birds in the 1870s, claimed that without environmental changes organisms did not evolve at all. He supported this claim by a close relationship between geographic variability and climatic differences and by the fact that archaic forms occurred in a constant environment of deep ocean (Allen 1906). Nevertheless, Cope and his views were more in the public eye, and Lamarckism with an emphasis on the internal forces of progressive evolution has come to be considered as a salient feature of the American Neo-Lamarckian school (Bowler 1983).

At the same time as the authors mentioned above, the laws of development and evolution were reflected upon by **Karl Ernst von Baer** (1792–1876), a Russian academician of Baltic German descent. Early in his scientific carrier, that is, long before Darwin's theory, Baer expressed deep scepticism towards the treatises of contemporary evolutionary philosophers, in particular, mocking Lamarck's idea of exercises. At the same time, the idea of evolution seemed to strike a chord with him though he hardly ever mentioned it in his writings. (Baer always expressed his views on the subject with the utmost caution, and for a long time they were known only to his inner circle.)

To note, Baer often verged to extremes in his choice of research topics. After his embryological discoveries met ill quarter from the scientific community, he vouched to ignore embryology for 9 years. He kept his word; in fact, he stayed away from this science for an even longer period. After the emergence of Darwin's theory Baer also decided to keep away from discussions on this topic allegedly remarking that it was unwise to become involved in heated debates at his age. (This does not sound convincing, however, as he continued to work in the field and to publish papers, including papers on fishing and fish management, which was a touchy topic in Russia at that time.) Finally, however, "a desire to uphold his principles got the better of senile prudence" (Raykov 1961, p. 397). Though he had lost his sight, in the

1870s Baer wrote (or, rather, dictated) a book about evolution, finding in himself enough force to generalise an extensive material and to analyse the evolutionary theory.

In this work (von Baer 1876) Baer expressed a number of objections against Darwinism that were typical of his time: there are no transitive forms between the main types of the animal world, the summation of fine changes can only yield a chaos of uncertain forms, variations lacking artificial support disappear and altered organisms revert to the initial condition etc. In particular, he objected to the idea of a boundless non-directed variability (von Baer 1876; Raykov 1961). For this reason, Baer is sometimes named among the forerunners of the theory of directed evolution (Rensch 1954; Riedl 1975) though the actual contribution of his writings to the formation of this theory seems insignificant. Firstly, they were, in essence, philosophic speculations, an already obsolete genre in the 1870s. Baer was unenthusiastic and condescending about evolution in general. He claimed to have arrived at similar conclusions before Darwin and disagreed with him at the same time. He confused the problem of evolutionary factors and that of the origin of species, being especially displeased with the idea about the descent of man from animals. Secondly, the last studies of Baer passed largely unnoticed. They were written in German, published in Saint Petersburg and printed using the Gothic font. As a result, their readership was restricted both in Russia and outside it. Thirdly, Baer strongly criticised the idea of recapitulation, while most advocates of orthogenesis supported it. Finally, Baer paid much attention to the analysis of the purposes of development. He used the notion *Zielstriebigkeit* to designate the driving force of development, meaning the ability to move towards a certain purpose. Though he mostly discussed embryological development, he expressed similar ideas in evolutionary context (*zielstriebigen Weltschopfung* – purposeful creation of the world; von Baer 1876. S. 237). The supporters of orthogenesis usually expressed opposite views.

The next remarkable event in the history of the concept of directed evolution was the publication of Nägeli's voluminous treatise *Mechanical-physiological theory of evolution* (1884). Unlike his first brief papers, it was presented not as an addition to Darwin's theory but as an alternative to it. Nägeli resolutely denied the principle of natural selection. His central idea was the existence of idioplasma, a substance contained in the entire organism which determined its hereditary properties and could transform spontaneously in the series of generations. Nägeli characterised this capacity as a perfective force (*Vervollkommungskraft*), that is, thought that evolutionary transformation was mainly progressive. As before, Nägeli criticised the principle of usefulness. His major argument was that many features of plants were not adaptive; moreover, it was the non-adaptive characters that distinguished high-level taxa (symmetry types of flowers, phyllotaxis, the position of veins etc.). Nägeli claimed that evolution was not synonymous with adaptation and that the main evolutionary trends did not depend on the environment. He admitted that some characters might be useful and polemicized with Kölliker on this matter.

Nägeli's concept stands out among the first works on directed evolution because he was well-versed (for his time) on the subject of mechanisms of heredity. Partly that was due to the fact that he was familiar with the studies of Gregor Mendel.

Nägeli was one of those prominent scientists to whom Mendel wrote letters describing the results of his studies and very nearly the only one who deigned to reply. On the whole, Nägeli disapproved of the "pea experiments" and hardly ever cited Mendel. Nevertheless, through his contact with Mendel he became close to the beginnings of genetics (Gaysinovich 1988).

The blunder with Mendel notwithstanding, Nägeli appeared to be perspicacious. Several of his insights turned out to be correct. For instance, he claimed that heredity was determined by a substance consisting of groups of crystal molecules and possessing a combination of properties (Nägeli 1884, p. 23). These *Micellen* suggest something filamentous, and it is tempting to believe that Nägeli anticipated DNA. It is possible that the ideas about some hereditary substance were born from Nägeli's observations of crystals and other small structures in cells.

Despite Nägeli's familiarity with the incipient genetics, his concept has often been considered as Lamarckian (e.g. Schmalhausen 1969). This was probably due to certain contradictions in his speculations about the direct impact of the environment. Nägeli believed that random changes were possible only under the external impact and that these changes were negative rather than positive (Nägeli 1884, p. 12). Since, according to Nägeli, organisms changed in certain directions, one might conclude that they were driven by the environment. However, Nägeli was sceptical about the inheritance of acquired characters. In his opinion, direct external influence could cause changes in the hereditary substance only in singular instances and in case of a prolonged impact while spontaneous transformation of idioplasma was the main mechanism of evolution.

One can also find inklings of the idea about the separation of macro- and microevolution in Nägeli's concept. Indeed, Nägeli distinguished two groups of characters: physiological characters, which change under the influence of natural selection, and morphological characters, which change according to the "principle of perfection" (Junker 2004, p. 357). Such comments indicate the synthetic composition of Nägeli's concept. It united major non-Darwinian evolutionary trends of the time into a coherent whole. Even some staunch supporters of Darwinism appreciated his work. For instance, **Nikolai A. Kholodkovsky** (1858–1921), after some scorching criticism of Nägeli's theory, noted that "there is no denying the consistency and fascination of this theory. However strange and arbitrary might the hypothesis of micellar structure and the plethora of additional hypotheses seem, all these speculations are so intricately connected with each other and so well adjusted to explanation of several facts that in end the hypothesis looks much more plausible than at first sight" (Kholodkovsky 1888 cited in Kholodkovsky 1923, p. 104). Kholodkovsky attached a particular value to Nägeli's idea about hereditary substance, remarking that: "the idea of idioplasma is a great idea indeed" (Ibid., p. 113).

Nägeli's ideas matched most palaeontological studies of the time. In the late nineteenth century palaeontologists such as **Charles Depéret** (1854–1929), **Wilhelm Heinrich Waagen** (1841–1900), **Melchior Neumayr** (1845–1890) and others often expressed more or less articulate ideas about directed evolution. They searched for regularities, laws and internal plans in phylogenetic series and usually managed to find something. In some cases they merely presented abstract speculations. A French palaeontologist **Albert Gaudry** (1827–1908), who greatly contributed to the promotion of evolutionary theory in France, thought that natural history developed according to a definite plan but always noted that the mechanisms of evolution were so poorly known that further discussions were meaningless (Gaudry 1896).

Some other authors analysed various aspects of directed evolution in more detail or described the results of their research supporting the idea of orthogenesis. In addition to Cope's laws, several similar generalisations were made such as the law of decreasing variability suggested by Daniele Rosa, the law of increasing body size suggested by Charles Depéret etc. Another idea that emerged from paleontological data at that time was that of inertia in evolution. **Ludwig Heinrich Philipp Döderlein** (1855–1936) suggested that evolution went on in a certain direction even after the maximal benefit had been passed, citing series of evolutionary development of felids, cervids and an extinct mammal family Oreodontidae as examples. According to Döderlein, *vis inertia* showed that not everything in evolution could be explained by natural selection and struggle for existence (Döderlein 1888).

By the last decade of the nineteenth century internally predetermined evolution independent of natural selection had been discussed in numerous works and the term *orthogenesis* had been coined. At least three scientists are credited with its invention: Theodor Eimer (Belgovsky and Khvostova 1948), Ernst Koken (Guyénot 1930) and Wilhelm Haacke (Rensch 1954). Such conflicting reports are unsurprising given that the nineteenth century authors wrote voluminous treatises often coining multitudes of terms. I consider it most likely that the term orthogenesis was coined by the German zoologist **Wilhelm Haacke** (1855–1912). At least, I have not come across a single mention of this term before 1893, when Haacke's book on heredity was published (Haacke 1893a). In addition, Eimer (1897) also considered this book as the original source of the term.

Wilhelm Haacke was a student of Ernst Haeckel. Similarly to his teacher, he wrote several treatises on general problems of biology. He was known among zoologists for his studies in Australia, where he was the first to describe the egg of the echidna. His experimental studies on heredity also attracted the attention of specialists. Crossing mice, he came close to a rediscovery of Mendel's laws (Schmidt 2000). In the book *Origin of forms and heredity* (1893a), Haacke put forward a question whether evolutionary changes were ordered or not. For some reason, however, he reformulated it in his own language presenting it as a choice between amphigenesis and orthogenesis. The former term was hardly ever used since, even by Haacke himself, who came to believe in orthogenesis. Haacke compared evolution with the movement of planets, which followed Kepler's laws, and with other phenomena subject to the laws of nature.

Orthogenesis in Haacke's understanding was closely related with his theory of heredity. It was similar to that by Weismann, who suggested the existence of units of heredity ("determinants", "ides", "idants") joining each other in various combinations. Haacke thought that these units were polyhedrons. The number of possible combinations of polyhedrons is limited by their shape, which leads to a predisposition to certain variants of organisms' organisation, i. e. orthogenesis. Haacke called

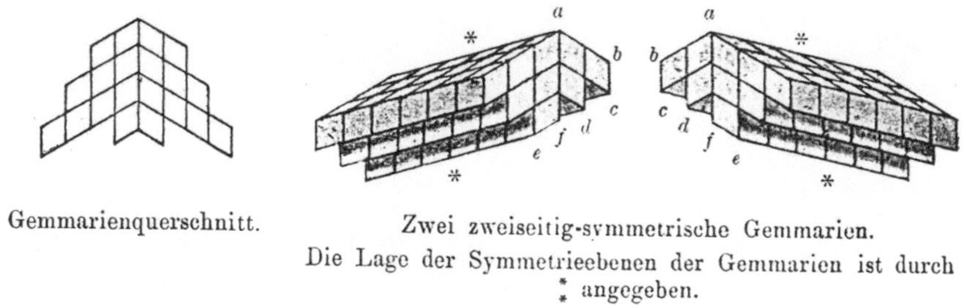

Fig. 2.1 Gemmae and Gemmaria of Wilhelm Haacke

these particles *Gemmae* and their complexes, *Gemmaria*. He specially noted that his purpose in coining new terms was to avoid confusion with other theories. He admitted that his theory might seem similar to that by Weismann, who also postulated that *Iden* had a definite form. Yet Haacke believed in the originality of his own concept. The doctrine of *Gemmaria* addressed, in his opinion, a much broader scope of phenomena. It aimed to explain not only heredity but the entire variety of evolutionary and developmental processes such as the change of symmetry types, the origin of various *Gemmaria* from an initial *Urform* and other enigmatic natural processes. He thought that the rhomboid form would be most suitable for *Gemmae* because both bilateral and radial structures could easily form and transform from such building blocks (Fig. 2.1). Haacke claimed that the study of units of heredity was a geometric rather than a phylogenetic task. He even argued about this with Carl Nägeli, who, it would seem, was his predecessor in the development of the orthogenesis concept.

The naïve idea of *Gemmae* as building blocks never became popular. In general, Haacke's evolutionary ideas were superficial and contradictory. While some of his speculations were similar to those of Weismann, his passionate writings on habits, exercises and the action of environment were worthy of Lamarck (Haacke 1893b).

However, one of Haacke's terms, *orthogenesis*, gradually found its way into biology. This happened mainly owing to a Swiss-German zoologist **Theodor Eimer** (1843–1898). His book *Orthogenesis of butterflies* (Eimer 1897) is a much more solid treatise than the "popular science" of Haacke. Eimer conducted several series of studies on different animals such as lizards, butterflies and foraminiferans. Early in his scientific carrier he worked with Weismann and Kölliker, and many of his evolutionary ideas seem to have arisen from the scientific antagonism between these two personalities (Fickert 1899; Bowler 1979).

In *Orthogenesis of butterflies* Eimer analysed coloration in different butterfly species (especially swallowtails), applying the term "orthogenesis" to differences between species rather than variation within a species. He described gradual series of transformations formed by the coloration patterns and believed them to display evolution. Critics pointed out that all these butterflies were extant and these series did not indicate that the same transformations had occurred in the past. They also noted that Eimer analysed single characters rather than the entire organism (e.g. Rabaud 1953). Eimer, however, did not seem in the least perturbed by this. He

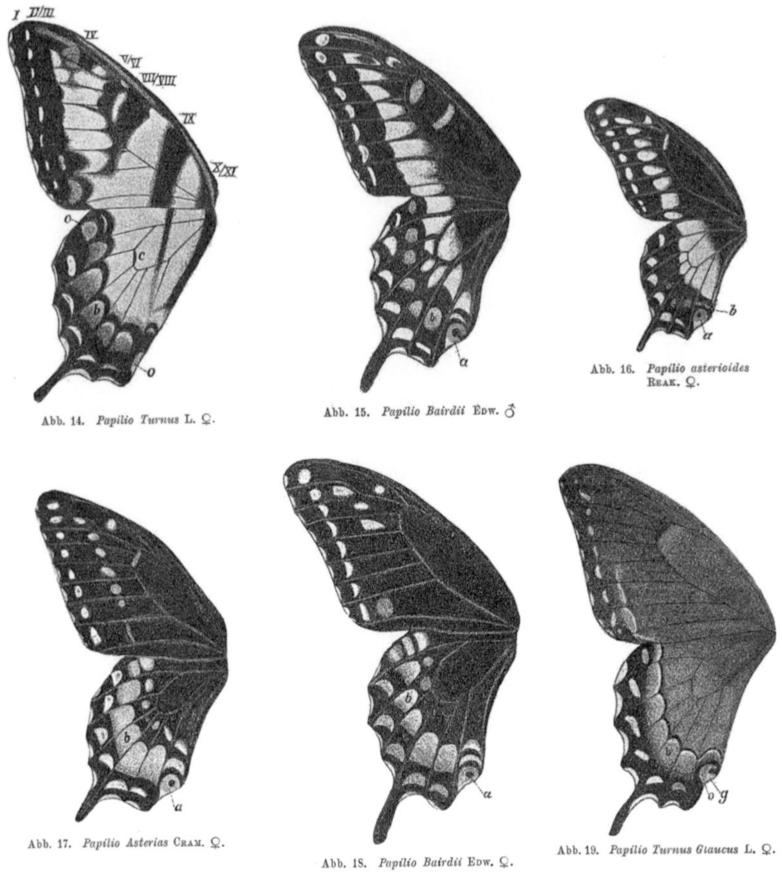

Fig. 2.2 Orthogenesis of butterflies (Eimer 1897)

considered orthogenesis of butterflies as a manifestation of general evolutionary laws and was quite sure about which kinds of coloration were primitive and which were secondary. For instance, he claimed that coloration in swallowtails evolved from complex patterns to more plain ones (Fig. 2.2). This process, it seemed, defied selection and adaptation.

Interestingly, *Orthogenesis of butterflies* was the second part of the treatise *The origin of species as a result of inheritance of acquired characters according to the laws of organic growth* (Eimer 1888).[2] This sounds entirely different from orthogenesis (at least in the best-known meaning of this term) and even much more different from the ideas of Weismann and Haake. Indeed, Eimer aimed to criticise the

[2] *Die Entstehung der Arten auf Grund von Vererben erworbener Eigenschaften nach den Gesetzen organischen Wachsens* (Eimer 1888). English translation by Joseph Thomas Cunningham as *Organic Evolution as the result of the inheritance of acquired characters according to the laws of organic growth*, Macmillan & Co., London, 1890.

former; his later book was even subtitled *Objections to Weismann.* In this case, it would seem that Eimer was close to American Neo-Lamarckians, and actually they did cite Eimer approvingly (Cope 1890). For instance, Cope agreed with Eimer's suggestion that variations of coloration were not random but followed certain regularities, noting that he came to the same conclusions studying the coloration of snakes. Cope also supported Eimer's criticism of Weismann (Cope 1890). Nevertheless, Eimer differed from his American colleagues at least in one significant point: he disagreed with the concept of exercises. In his opinion, exercises, similarly to natural selection, could only cause minor deviations from directed evolutionary process; this direction did not emerge spontaneously but was determined by a direct influence of the environment, given that organisms were predisposed to variation in certain direction. According to Eimer, such a response of the organism to a direct external impact was not adaptation. Eimer was a vociferous critic of ideas suggesting adaptation, especially scorning mimicry.

Strangely, Eimer was also dissatisfied with the ideas of Carl Nägeli. He thought that the notion of idioplasma smacked of mysticism. In his opinion, Nägeli's "perfective force" was immaterial while the action of environment was tangible. That was why Eimer tried to reconcile spontaneous transformation under the influence of internal forces with the direct action of the environment. He was especially consistent in his attempts to analyse evolution by analogy with the growth of crystals. The environment influences this process by providing the building material but the shape of crystals is determined by their chemical composition. Eimer believed that it was the same with evolution: it was triggered by the environment but mostly determined by the ability of organisms to change in certain directions.

Though Eimer's views were quite different from those of other pioneers of directed evolution, his *Orthogenesis* was an important milestone in the history of this idea. In the 1890s it was no longer scattered across the works of handful of authors: some results had been summed up and a term emerged to be used outside the works of its author. In addition to the book of Eimer, the last generalising books by Cope (1896) and Hyatt (1897) were published at about the same time. They promulgated directed evolution and criticised Darwin's theory. Taking all this into account, we can consider that the theory of directed evolution was formed or, rather, established in 1897.

In the course of the entire twentieth century, the followers of orthogenesis (as well as those of other biological heresies) were repeatedly accused of poor knowledge of Darwin's works. It would seem that Darwin had given answers to all questions and his critics simply failed to read or understand him properly. At the same time, it was not uncommon for the supporters of orthogenesis to consider Darwin's works as the Holy Writ of biology and to try to find hints about orthogenesis there. That is why, in order to get a coherent picture of the formation of orthogenesis in the nineteenth century, one should address Darwin's heritage.

The main arguments in favour of orthogenesis were suggested, as we have seen, in the 1860s. Darwin was still actively working at that time. What did he say in reply to these arguments? Strictly speaking, almost nothing. He was much more interested in other problems and could not address all the objections he received. After

The Origin of Species had been published, Darwin began to collect reviews of his book but, having collected 265, stopped in desperation (Darwin 1887). He did anticipate the major critical remarks, addressing them in his writings as best he could. Being a great naturalist, he could not ignore the facts and arguments in favour of directed evolution. Usually, however, he did not discuss the works where they were presented, Nägeli's article being an exception. In the sixth edition of *The Origin of species* (1872) he mentioned an "admirable essay" by Carl Nägeli. Darwin admitted the seriousness of the argument about the existence of neutral characters distinguishing plant families such as numerical divisions of the flower parts, the position of ovules, the shape of seeds etc.). "Nevertheless, we ought, in the first place, to be extremely cautious in pretending to decide what structures now are, or have formerly been, of use to each species. In the second place, it should always be borne in mind that when one part is modified, so will be other parts, through certain dimly seen causes, such as an increased or diminished flow of nutriment to a part, mutual pressure, an early developed part affecting one subsequently developed, and so forth,—as well as through other causes which lead to the many mysterious cases of correlation, which we do not in the least understand. These agencies may be all grouped together, for the sake of brevity, under the expression of the laws of growth. In the third place, we have to allow for the direct and definite action of changed conditions of life, and for so-called spontaneous variations, in which the nature of the conditions apparently plays a quite subordinate part." (Darwin 1872, p. 175).

We can see that the data and considerations presented by Nägeli could hardly be explained in terms of natural selection. That is why Darwin tried either cast doubt on them or appealed to our insufficient knowledge. Similarly to the authors of concepts concerning directed evolution, Darwin referred to some "mysterious laws of growth". Analysing facts presented by Nägeli, Darwin gave examples when characters of plants seemed to be neutral at first (different length of pistils and stamens in dimorphic and trimorphic plants) and were found be functional after further research (different role in the process of fertilization by insects). Darwin believed that these functions "have probably been gained through natural selection" (Ibid., p. 172). However Darwin admitted that "with plants many morphological changes may be attributed to the laws of growth and the inter-action of parts, independently of natural selection", for example, tetramerous or pentamerous structure of flower (Ibid., p. 175). He believed that these characters, though important for taxonomists, were unimportant for plants and so the overall importance of such differences for evolution was insignificant. Darwin's main conclusion on this issue was as follows: "Although we have no good evidence of the existence in organic beings of an innate tendency towards progressive development, yet this necessarily follows, as I have attempted to show in the fourth chapter, through the continued action of natural selection." (Ibid., p. 176). Later Darwin seem to have taken back even such small "concessions" to Nägeli concerning adaptation and came to believe that *all* characters could have originated due to adaptation through natural selection. He expressed this viewpoint in letters to his colleagues written in the 1880s (Takhtadzhyan 1991).

A critical attitude to adaptation is not the only aspect of the concepts of directed evolution. Their authors emphasised a regular pattern of variation believing that organisms varied in certain directions rather than randomly. Darwin addressed this topic several times, although not in the context of the analysis of directed evolution. He noted repeatedly that selection did not cause changes but used the material supplied by nature. At the same time, Darwin clearly did not want to analyse what exactly nature supplied selection with, and tried to hush up this question. Finally, he concluded that variation was secondary as compared to natural selection. At the same time, he also discussed some data indicating a regular nature of variation and the existence of certain constraints. He mentioned B. D. Walsh's law of *Equable Variability*: "if any given character is very variable in one species of a group, it will tend to be variable in allied species; and if any given character is perfectly constant in one species of a group, it will tend to be constant in allied species" (Darwin 1875, p. 344). This observation was probably quite well-known since similar statements can be found in other sources of that time and were considered as an indication of the existence of certain directions in variation. For example, Joseph Duval-Jouve described parallel variations within several plant genera (*Junkus, Luzula, Vulpia, Bromus* etc.) and came to the conclusion that though "non-variation and absolute constancy of species is a not applicable hypothesis, this does not mean the other extreme, i. e. unlimited and fortuitous variability, which would be a denial of the species idea" (Duval-Jouve 1865, p. 209). Darwin believed in this very "extreme". He considered Walsh's law as a demonstration of a great variability of those characters interesting for human needs but not as an indication of constraints on variation.

Another law discussed by Darwin is a "law of analogous variation": "this law implies, as stated in a previous chapter, that the varieties of one species frequently mock distinct but allied species" (Darwin 1875, p. 255). However, even this law did not inspire Darwin to analyse definite directions of variations. Darwin considered it merely as an indication of the common origin of various species: "this fact is explained, according to the views which I maintain, on the principle of allied species having descended from one primitive form" (Ibid.). Darwin thought that variability was virtually unlimited and could always supply natural selection with some resources. He mostly based this claim on the human practice of selection, having collected abundant examples of a successful creation of new breeds. He quoted approvingly the words of Lord Somerville about sheep-breeders: "It would seem as if they had chalked out upon a wall a form perfect in itself, and then had given it existence" (Darwin 1987, reprint of the first edition [1859], p. 90). However, he had to admit that sometimes, when selecting new breeds, humans wished for the impossible. When English farmers decided to get cows with thick hams, they soon abandoned this attempt since such cows perished too often during delivery. Evidently such cases suggested the existence of limitations on variability. Darwin wrote: "Will a gooseberry ever weigh more than that produced by 'London' in 1852? Will the beet-root in France yield a greater percentage of sugar? Will future varieties of wheat and other grain produce heavier crops than our present varieties? These questions cannot be positively answered; but it is certain that we ought to be cautious in answering them by a negative." (Darwin 1883, p. 228). Though he admitted the existence of some limitations, they were, in his opinion, too weak to influence the success of selection and, thus, evolution. He apparently understood that it was difficult to prove this viewpoint and so pointed that it was impossible to prove or disprove the alternative ones: "the ordinary belief that the amount of possible variation is a strictly limited quantity, is likewise a simple assumption." (Darwin 1872, p. 82). (This sentence was added in the 6th edition of *The Origin of Species*.) "Unknown laws of growth," which could cause the presence of non-adaptive characters in organisms, were analysed by Darwin in a similar manner. He believed that selection could do everything by small steps over a long period, comparing natural selection with an architect who is forced to build a majestic building with crude stones and yet fulfils this task successfully.

Darwin's viewpoint on constraints seems to follow logically from his theory. He probably realised that to pay attention to restrictions of variability would mean to admit that these constraints directed evolution. This could not be known and was therefore unacceptable. "Fluctuations"—minute distinctions between closely related organisms or their parts—could always be revealed and explained by small distinctions in environmental influences. For example, seeds in a boll are slightly different because of the differences in their position respectively of nutritive canals: seeds get different amounts of nutrition and become different. Sources of such fluctuations can be found and explained. On the contrary, the sources of *constraints* to variation are obscure. Therefore, Darwin tried to avoid this topic and did so quite successfully. The limits of variation are not mentioned even in a special study analysing Darwin's ideas on variation (Winther 2000).

Darwin collected abundant facts concerning variability and discussed them carefully but he also placed the discussion on the properties of variation so that it would not take the reader by surprise. He noted repeatedly that selection itself did not cause changes and that it was based on the material provided by "nature". He evidently did not wish to see the analysis of this "nature" at the foreground of his theory but wanted to postpone it for the future.

It is known that Darwin prepared his famous book *On the Origin of Species* (1859) rather hastily as a brief version of his theory after receiving a letter and a manuscript by Alfred Wallace, who had elaborated a similar concept. Originally Darwin had planned to write a "big species book" which would cover a wide range of topics. The analysis of the limits of variation was not included into this "brief version". After its publication Darwin finished this work but presented the additional parts as separate treatises on domestication. He put the chapter about the properties of variation in their end. Such a layout seemed more convenient though Darwin himself remarked that it would have been better to begin from this topic. In the chapter on the natural selection he wrote the following: "Hence our discussion on the causes and laws of variability ought in strict order to have preceded the present subject, as well as inheritance, crossing, etc.; but practically the present arrangement has been found the most convenient" (Darwin 1883, p. 176). "The present arrangement" in this case means placing numerous facts concerning selection and heredity in the beginning, and some ideas about the nature of variability closer to the end. This topic, in fact, occupies meagre two pages in all these treatises.

Another important point of difference between Darwin and most followers of directed evolution concerns the role of spontaneous transformation in evolution. Do organisms change in a constant environment? Darwin answered this question in the negative. He believed that any change was due to the impact of environment, either on the organism itself or on its ancestors.

Darwin gave examples of "sports" (somatic mutations of plants) and other instances of sudden emergence of new variations but tried to explain them by the impact of the environment on the ancestors, considered them insignificant or simply stated that the reasons were obscure. Now that we are more familiar with the molecular mechanisms of heredity, we realise that replication, reparation, mutation etc. are incredibly complex processes and errors are inevitable. This is why variability is an essential feature of the living world. It would be unfair to blame Darwin for not understanding this. At the same time, there seems to be a mistake in his reasoning even by the standards of his times. Looking for the source of variation in the action of environment, Darwin came close to what we call Lamarckism. For instance, he did not doubt the phenomenon of telegony, that is, the "direct action of male element" on the heredity of the female. He cited an example when "a nearly purely-bred Arabian chestnut mare bore a hybrid to a quagga; she was subsequently sent to Sir Gore Ouseley, and produced two colts by a black Arabian horse. These colts were partially dun-coloured, and were striped on the legs more plainly than the real hybrid, or even than the quagga. One of the two colts had its neck and some other parts of its body plainly marked with stripes. Stripes on the body, not to mention those on the legs, are extremely rare,—I speak after having long attended to the subject,—with horses of all kinds in Europe, and are almost unknown in the case of Arabians. But what makes the case still more striking is that in these colts the hair of the mane resembled that of the quagga, being short, stiff, and upright. Hence there can be no doubt that the quagga affected the character of the offspring subsequently begot by the black Arabian horse." (Darwin 1875, p. 435).

In this case, according to Darwin, the mare acquired from the colt a new feature (striped coloration) for the offspring and passed it to the offspring from another male. Darwin also believed in the existence of "graft hybrids", referring to the results of experiments on potato. Grafting of white, black and red potatoes resulted in numerous intermediate varieties. Similar cases were described for many other plants (Darwin 1875; Ch. XI).

Darwin also analysed the role of exercises and the use or disuse of organs in the evolution of animals. For example, he found that the brain of domestic rabbits was smaller than that of wild ones. According to Darwin, when we remember that domestic rabbits "from having been domesticated and closely confined during many generations, cannot have exerted their intellect, instincts, senses, and voluntary movements, either in escaping from various dangers or in searching for food, we may conclude that their brains will have been feebly exercised, and consequently have suffered in development. We thus see that the most important and complicated organ in the whole organisation is subject to the law of decrease in size from disuse." (Darwin 1875, p. 135).

All this smacks of Lamarck but Darwin had a much more solid base referring to numerous results of empirical studies including some of his own. His observations on heredity resulted in a “provisional hypothesis of pangenesis” (which did turn out to be provisional). According to this hypothesis, each cell contains small particles called gemmules, which correspond to the cell in respect to heredity. They circulate in the organism and penetrate genital tissues. The action of the environment changes somatic cells together with the gemmules, and the latter pass new hereditary properties to genital cells. The hypothesis had few followers and Darwin did not promote it too much himself though he did not want to abandon it either. It disagreed with the results of experiments made by Francis Galton (1822–1911), who transfused blood from black rabbits to white ones and vice versa but did not see any influence of this process on the offspring.

It seems that as concerns the nature of variation and heredity Darwin came to a dead end. Nevertheless, his works were always solid. He took care to back up every claim he made with arguments, and even thought it necessary to justify his very right to suggest hypotheses: “In scientific investigations it is permitted to invent any hypothesis, and if it explains various large and independent classes of facts it rises to the rank of a well-grounded theory” (Darwin 1875, p. 9). That is why his treatises were to become a biological Bible, where anyone was free to look for anything one might need. Many authors claimed that their theories were in some way anticipated by Darwin, and even such an exotic concept as symbiogenesis was presented as a new version of Darwin’s theory (Kozo-Polyansky 1921, p. 24). Orthogenesis seems to be a notable exception in this respect. Darwin never made any concessions to it. His ideas were antipodal.

Interestingly, both Darwin’s theory and orthogenesis have been accused of being teleological and generally non-materialistic. Some of Darwin’s statements do invite this kind of criticism. For example, he wrote that the giraffe was an animal “of which the whole structure is admirably co-ordinated for certain purposes, it has been supposed that all the parts must have been simultaneously modified; and it has been argued that, on the principle of natural selection, this is scarcely possible” (Darwin 1875, p. 206).

At the same time, Darwin, similarly to the authors of orthogenetic concepts, was careful to distance himself from anything that might have seemed non-scientific. In trying to do so, he pointed (again, similarly to his opponents) to an analogy between evolutionary mechanisms and processes studied by other natural sciences. “No one objects to chemists speaking of ‘elective affinity;’ and certainly an acid has no more choice in combining with a base, than the conditions of life have in determining whether or not a new form be selected or preserved. The term is so far a good one as it brings into connection the production of domestic races by man’s power of selection, and the natural preservation of varieties and species in a state of nature. For brevity sake I sometimes speak of natural selection as an intelligent power;—in the same way as astronomers speak of the attraction of gravity as ruling the movements of the planets, or as agriculturists speak of man making domestic races by his power of selection. In the one case, as in the other, selection does nothing without variability, and this depends in some manner on the action of the surrounding circumstances on the organism. I have, also, often personified the word Nature; for I have found it difficult to avoid this ambiguity; but I mean by nature only the aggregate action and product of many natural laws,—and by laws only the ascertained sequence of events." (Darwin 1875, p. 7).

This is as close to a rapprochement between orthogenesis and Darwinism as we can get. However, the only thing they had in common was a desire to explain evolution scientifically. In all essential points, orthogenesis is an alternative to Darwin's theory.

Unfortunately, it is difficult to find the first definition of this alternative or to name its author. Orthogenesis was born from a tangle of various, sometimes opposing biological ideas. Its roots intertwine with Lamarckism, Neo-Lamarckism, incipient genetics, the Neo-Darwinism of Weismann and even the anti-evolutionism of Cuvier.

In the second half of the nineteenth century the concept of directed evolution was developed by Albert von Kölliker, Carl Nägeli, Edward Cope, Alpheus Hyatt, Wilhelm Haacke and Theodor Eimer. These authors worked more or less independently of each other, and each can be credited with being in some respect the pioneer of orthogenesis—one wrote the first article on the topic, another was the first to present it as a new theory, yet another analysed some significant aspect. Even if we approach to the issue of priority in the formal way, this would not clarify the situation much. Nägeli presented his ideas about evolution as early as in 1856 but the idea of directed evolution carries weight only in comparison with Darwin's theory. So it is the second publication of Nägeli on this topic (1865) that should be considered as the first presentation of his version of an evolutionary theory. In between Nägeli's two papers, the concept of directed evolution was presented by Kölliker (1864), who can thus also be considered as its founder.

However, priority does not seem to have been an issue at that time. Although many of the authors mentioned above cited each other's works and some, such as Nägeli and Kölliker, even knew each other well personally, they were generally uninterested in the studies of their colleagues. They never attempted a synthesis of the ideas concerning directed evolution. Each author presented his doctrine using his own terms and generally his own scientific language and was apparently unconcerned by the doctrines of others.

To put it shortly, the first works dealing with directed evolution were disconnected. That is why we cannot find in them the first wording of this concept, however hard we might try. This elusive concept can be deduced from the context of the writings of several authors, who claimed that organisms changed spontaneously in definite directions and denied the leading role of natural selection. As for the rest, their ideas had little in common.

The term "orthogenesis" was ambiguous from the very start. It was used in relation with antipodal views on heredity and variation. At first mention, orthogenesis concerned variability rather than evolution but it soon came to be used for various phenomena associated with directed evolution such as differences between species, evolutionary tendencies of higher taxonomic groups, predetermined evolution etc. Priority is often considered as a means to maintain at least a simulacrum of order in biology but in this case it does not help much. It is only with numerous qualifications that the term "orthogenesis" can be used to characterise all the concepts concerning directed evolution.

References

Allen J (1906) The influence of physical conditions in the genesis of species. Annual report of Smithsonian Institution (1905), Washington, p 375–403
Belgovsky ML, Khvostova VV (1948) Explanation of some terms (Poyasnenie nekotorykh terminov). In: Russian edition: Simpson GG (1948) Tempy i formy evolyutsii. Gosudarstvennoe izdatel'stvo inostrannoy literatury, Moscow, p 334–340
Bowler P (1979) Theodor Eimer and orthogenesis. J Hist Med 34:40–78
Bowler PJ (1983) The eclipse of Darwinism. Johns Hopkins Univ Press, Baltimore
Bowler PJ (1989) Evolution. The history of an idea. Univ of California Press, Berkeley
Brocchi G (1843) Fossil subapennine conchiology, with geological observations on Apennines and adjacent areas (Conchiologia fossile subapennina, con observazioni geologiche sugli Appennini e sul suolo adiacente): In 2 vol. V 1. Giovanni silvestri, Milano
Cope ED (1868) On the origin of genera. Proceedings of academy of natural sciences of Philadelphia 4:242–305
Cope ED (1890) Eimer on evolution. Am Nat 5(24):751–754
Cope ED (1896) Primary factors of organic evolution. Open Court Publishing, Chicago
Cuvier G (1850) A Discourse on the revolutions of the globe, with notes and an appendix, according to recent studies (Discours sur les révolutions du globe, avec les notes et un appendice, d'après les travaux récentes). Libraire de Firmin Didot Frères, Paris
Dalcq A (1954) Ontomutations at the origin of mammals (Les ontomutations a l'origine des mammiferes). Bull Soc Zool Fr 79:240–255
Darwin Ch (1859 [1987]) The Origin of Species. (Reprint of 1st edition 1859) Suffolk: Penguin books
Darwin ChR (1872) The origin of species by means of natural selection, or the preservation of favoured races in the struggle for life. John Murray, London. 6th edition, with additions and corrections
Darwin ChR (1875) The variation of animals and plants under domestication. John Murray, London 2nd edn Vol 1
Darwin C (1883) The variation of animals and plants under domestication. Second Edition, Revised. D. Appleton, New York
Darwin ChR (1887) The life and letters of Charles Darwin, including an autobiographical chapter, John Murray, London
Definite vs. Fortuitous Variation in Animals and Plants (1892) Meehan Th, McMurrich JP, Allen JA, Cope ED (eds) Meeting of the American Society of Naturalists, Philadelphia, 1891. Am Nat 26: 83–92
Devillers Ch, Chaline J (1993) Evolution. An evolving theory. Springer Verlag, Berlin/Heidelberg
Döderlein L (1888) Phylogenetic observations (Phylogenetische Betrachtungen). Biologisches Centralblatt 7:394–402
Duval-Jouve M (1865) Parallel variations of congeneric types (Variations parallèles des types congénères). Bulletin de la société botanique de France 12. Comptes rendus de séances 3:196–211
Eimer Th (1888) Die Entstehung der Arten auf Grund von Vererben erworbener Eigenschaften nach den Gesetzen organischen Wachsens. I Teil. Fischer, Jena. English edition: Eimer (1890) Organic Evolution as the result of the inheritance of acquired characters according to the laws of organic growth (trans J Th Cunningham), Macmillan & Co., London

Eimer Th (1897) Orthogenesis of butterflies. An argument determines directed development and the impotence of natural selection for the origin of species. At the same time a rejoindre to August Weismann (Orthogenesis der Schmetterlinge. Ein Beweis bestimmt gerichteter Entwickelung und Ohnmacht der naturlichen Zuchtwahl bei der Artbildung. Zugleich eine Erwiderung an August Weismann). W. Engelmann, Leipzig

Fickert C (1899) Gustav Heinrich Theodor Eimer. Tuebinger zoologisher Arbeiten. III Band 1898-1899. Verlag von Wilhelm Engelmann, Leipzig, p 5–11

Gaudry A (1896) An essay in philosophical palaeontology (Essai de paleontology philosophique). Masson et C-ie, Paris

Gaysinovich AE (1988) Origin and development of genetics (Zarozhdenie i razvitie genetiki). Nauka, Moscow

Gutmann WF (1996) Are there alternative ways for the development of the organisation of living organisms I (Gibt es Alternativwege fuer die Entwicklung der organisation von Lebewesen I). Nat Mus 126(8):250–262

Guyénot E (1930) Variation and evolution (La variation et l'évolution). G. Doin et C-ie, Paris

Haacke W (1893a) Form and inheritance. Developmental mechanichs of organisms (Gestaltung und Vererbung. Eine Entwickelungsmechanik der Organismen). T. O. Weigel Nachfolger, Leipzig

Haacke W (1893b) The creation of the animal world (Die Schöpfung der Tierwelt). Bibliographisches Institut, Leipzig

Hyatt A (1866) On the parallelism between the different stages of the life in the individual and those in the entire group of the molluscous order Tetrabranchiata. Memoirs read before the Boston Society of Natural History, being a new series of the Boston Journal of Natural History 1: 198–210

Hyatt A (1897) Cycle in the life of individual (ontogeny) and the evolution of its own group (phylogeny). Proceedings of the American Academy of Arts and Sciences 32(10): 207–224

Junker Th (2004) The second Darwinian revolution. The history of synthetic Darwinism in Germany from 1925 to 1950 (Die zweite Darwinistische Revolution. Geschichte des Synthetischen Darwinismus in Deutschland 1925 bis 1950). Basilisken Presse, Marburg

Kanaev II (2000) Selected works on the history of science (Izbrannye trudy po istorii nauki). Aleteiya, St Petersburg

Kholodkovsky NA (1888) cited in Kholodkovsky NA (1923) Biological essays (Biologicheskie ocherki). In: Kholodkovsky NA, Pavlovsky EN (eds) A collection of selected articles on the theory of evolution and various problems of biology (Sbornik izbrannykh statey po teorii evolyutsii i razlichnym voprosam biologii). Posthumous edition. Gosudarstvenne izdatel'stvo, Moscow/Petrograd

Kölliker RA (1864) Über die Darwin'sche Schöpfungstheorie (Vortrag 1864 in Würzburg). Z Wiss Zool 14:174–186

Kölliker RA (1872) Morphology and developmental history of Pennatulid genera alongside general considerations concerning evolutionary theory (Morphologie und Entwickelungsgeschichte des Pennatulidenstammes nebst allgemeinen Betrachtungen zur Descendenzlehre). Christian Winter, Frankfurt a. M

Korzhinsky SI (1899) Heterogenesis and evolution. Towards the theory of the origin of species. I. (Geterogenezis i evolyutsiya. K teorii proiskhozhdeniya vidov. I.). Zapiski Imperatorskoy Akademii nauk, Seriya 8, 9(2): 1–94

Kozo-Polyansky BM (1921) Symbiogenesis in the evolution of plant world (Simbiogenezis v evolyutsii rastitel'nogo mira). Vestnik opytnogo dela 4:1–24

Kryzhanovsky SG (1939) The principle of recapitulation and the conditions of historical understanding of development (an essay of the theory of historical heterogenesis) (Printsip rekapitulyatsii i usloviya istoricheskogo ponimaniya razvitiya [ocherk teorii istoricheskogo geterogenezisa]). In: In memory of academician A. N. Severtsov (Pamyati akademika A. N. Severtsova), v 1. AN SSSR, Moscow, Leningrad, p 281–383

Lamarck JB (1809) cited in Lamarck (1914) English translation by hugh samuel roger elliott

Lamarck JB (1822) Natural history of invertebrate animals (Histoire naturelle des animaux sans vertèbres). T. 6. Paris,Verdière, translation into English: http:.www.nndb.com/people/275/000057104/
Nägeli C (1856) Individuality in nature with regard to plant world (Die Individualität in der Natur mit Berücksichtigung des Pflanzenreichs). Akademishe Vortrage, II. Verlag von Meyer & Zeller, Zurich, p 171-212
Nägeli C (1865) Origin and concept of species in natural history (Entstehung und Begriff der Naturhistorischen Art), Im Verlage der koenigl Akademia, München
NägeliC(1884)Mechanistic-physiologicaltheoryofevolutionarytheory(Mechanischphysiologische Theorie der Abstammungslehre). Druck und Verlag von R. Oldenbourg, Munich and Leipzig
Pfeifer EJ (1965) The genesis of American Neolamarckism. Isis 56(184):156–167
Rabaud E (1953) Fortuity and the life of species (Le hasard et la vie des espèces). Flammarion, Paris
Raykov BE (1952) Russian evolutionry biologists before Darwin (Russkie biologi-evolyutsionisty do Darvina). AN SSSR, Moscow, Leningrad
Raykov BE (1961) Karl Baer. His life and works (Karl Ber. Ego zhizn' i trudy). AN SSSR IIET, Moscow, Leningrad
Rensch B (1954) New problems of the evolutionary theory: transspecies evolution (Neuere Probleme der Abstammungslehre: transspeciphische Evolution). Enke, Stuttgart
Richardson RC, Kane TC (1989) Orthogenesis and evolution in the nineteenth century. In: Nitecki M (ed) Evolutionary progress. Univ of Chicago press, Chicago and London, pp 149–169
Riedl R (1975) The order of living things (Die Ordnung des Lebendigen). Verlag Paul Parey, Hamburg and Berlin
Schmalhausen II (1969) The problems of Darwinism (Problemy darvinisma). Nauka, Leningrad
Schmidt I (2000) Concise biographies (Kurzbiographien). In: Jahn I (ed) History of biology (Geschichte der Biologie). 3 neubearbeitete Auflage. Spektrum Academischer Verlag, Berlin and Heidelberg, pp 755–1000
Steinmann G (1908) Geological foundations of the evolutionary theory (Die geologischen Grundlagen der Abstammungslehre). Engelmann, Leipzig
Takhtadzhyan AL (1991) Darwin and the modern theory of evolution (Darvin i sovremennaya teoriya evolyutsii). In: Russian edition: Darwin Ch (1991) The origin of species (Proiskhozhdenie vidov). Nauka, St Petersburg, p 489–523
von Baer KE (1876) On the Darwinian teaching (Über Darwinsche Lehre). In: Studien aus dem Gebiete der Naturwissenschaften. Verlag der Caiferlichen Hofbuchhandlung H. Schmissdorf, St Petersburg, p 235–479
Winther RG (2000) Darwin on variation and heredity. J Hist Biol 33:425–455

Chapter 3
Evolutionary Biology at the Turn of the Twentieth Century. New Concepts of Directed Evolution in the 1900–1930s

Abstract At the turn of the twentieth century evolutionary biology entered a new phase, which lasted until the 1930s. Numerous, and quite different evolutionary concepts existed in the 1900s–1930s, while none of them seemed to dominate. Some of them were associated with orthogenesis, which was at the peak of popularity in the 1910s–1920s. New concepts of directed evolution were born in Russia (nomogenesis, historical biogenetics), the USA (aristogenesis), Austria (apogenesis), France (allelogenesis), Italy (ologenesis). The authors of these concepts found a lot of new data relevant to directed evolution but also confused this field coining numerous terms and claiming anew what had already been known.

Keywords Nomogenesis · Historical biogenetics · Aristogenesis · Ologenesis · Allelogenesis · Apogenesis

The boundaries of historical periods tend to become blurred under close scrutiny. However, most historians agree that at the turn of the twentieth century evolutionary biology entered a new phase, which lasted until the 1930s (Zavadsky 1973; Bowler 1983). This marked change was mainly associated with the establishment and a rapid progress of a new discipline—genetics. The first geneticists were mainly critical of Darwinism. **Hugo de Vries** (1848–1935), one of those who rediscovered Mendel's laws, went down in the history as the author of saltationism (mutationism) concept emphasising the importance of sudden changes in evolution. **William Bateson** (1861–1926), one of the major theorists of genetics of his time, and **Thomas Hunt Morgan** (1866–1945), the author of the chromosome theory of heredity, expressed their ideas in terms of regressive evolution, explaining the emergence of mutations by the loss of genes. They considered evolution (or, at least, much of it) as a process of dissipation of a pre-existing potential.

Lucien Cuénot (1866–1951), one of the pioneers of genetics, suggested the concept of preadaptation, which focussed on the adaptation of the organism to its own structure rather than to the environment. He pointed out that organisms could not decide to adapt to a certain environment and start to evolve in the right direction. On the contrary, they experience some changes first and then try to adapt to them choosing a suitable environment. So, preadaptation is more important than

I. Popov, *Orthogenesis versus Darwinism*,
https://doi.org/10.1007/978-3-319-95144-7_3

adaptation in evolution. This view was at odds with Darwinism, which focused on adaptation to the environment. After suggesting the concept of preadaptation, Cuénot wrote numerous other treatises on evolution, also opposed to Darwinism. At the same time, he thought that the main evolutionary problems were still unsolved (Cuénot 1932).

Another French geneticist, **Emil Gyuénot** (1885–1963), also emphasised the difficulties encountered within the paradigm of adaptation. He pointed out that the structure of many organisms, usually considered as normal and adapted to certain environment, could be just as well considered as non-adapted, anomalous or even monstrous. The entire variety of tetrapods including humans could be described in terms of teratology since each species was anomalous as compared to related ones. For example, examining the structure of limbs in the light of this concept, we can see that organisms adapted to some kind of movement demonstrate such "abnormalities" as the loss of limbs (snakes), shortening of limbs and fusion of fingers (pinnipeds), reduction of the upper limbs (non-flying birds) etc. Similar cases in other tetrapods are considered as monstrosities (Guyénot 1930; Nazarov 1974, 1984).

The first geneticists usually considered the mutation theory as an antithesis of the theory of natural selection. They believed that Darwin's theory concerns the so-called "fluctuations" or "variations", small stochastic differences between similar characters. Being based on experimental data, genetic studies commanded great respect at that time. This is why the advance of genetics has been associated by historians with a crisis of Darwinism. Moreover, since Darwinism is sometimes considered synonymous with evolutionary biology, it is sometimes thought that the very idea of evolution in general suffered an eclipse at the turn of the twentieth century.

The real state of things was, of course, much more complicated than that. Scientists such as Julian Huxley, Kliment Timiriazev, Boris Kozo-Polyansky and many others promoted Darwin's ideas at the time of the "crisis" and apparently encountered no difficulties in interpreting new genetic discoveries in the light of Darwinism. Moreover, the opposition of genetics and natural selection theory has often been overstated. Though several pioneers of genetics did have a critical attitude to Darwinism, they still found room for natural selection in their concepts. Even Hugo de Vries, who laid down the basis of "genetic anti-Darwinism", considered natural selection as a significant evolutionary factor: while some species originating by mutations are eliminated by natural selection, others survive. In this case, the differences between Darwinism and "anti-Darwinism" concerned the operational level of natural selection and the rates of evolution.

In a recent study, Peter Bowler (2003) has offered a fresh view on the matter: Darwinism did not undergo a "crisis" at the turn of the century because it had never dominated before. It is interesting that this viewpoint should be suggested by Bowler who described the crisis in much detail in an earlier book, *The Eclipse of Darwinism* (Bowler 1983). Apparently, after its publication Bowler found that theories opposed to Darwinism had existed—and had been quite influential—both before and during the alleged crisis.

Nevertheless, non-Darwinian concepts apparently rose to prominence in the early twentieth century, so it is reasonable to assume that there was some progress in this direction. In addition to a "genetic anti-Darwinism", "relapses into Lamarckism" were noted. Émile Gyuénot even made a remark to the effect that in the late nineteenth century "the virus of Lamarckism" infected biology again (Guyénot 1921, p. 614). This virus still struggled with genetics in the early twentieth century, when experiments on transformation of one species into another by direct action of environment were popular. It was only in the 1920–1930s that this approach was discredited by the discovery of Paul Kammerer's fraud, Lysenkoism and other events.

Some other areas of biology also entered a new phase in the early twentieth century, which was bound to influence evolutionary studies. Its first two or three decades are referred to as the geographical period of taxonomy (Huxley 1944). Studies on geographic variability increased in number; new taxonomic principles and new classifications of lower taxonomic units were elaborated. The species could be considered as a system of geographic units, or a system of ecological units, or a system of various units, or else as something uniform or even non-existent in reality. In addition to the habitual word "species", numerous alternative terms were proposed. In this connection, the "geographic period" was also a period of mental "ferment" of taxonomy (Mayr 1982). The "fermentation of ideas" stimulated a variety of views on evolution. They are usually considered as versions of Lamarckism but this is inexact. "Lamarckism" concerned only the hypothesis that geographic races originated due to direct action of the environment without natural selection. The inheritance of features of geographic races remained an open question. The first experiments aimed at answering this question only intensified the "mental fermentation" since both hereditary and non-hereditary characters of geographic races were found. These data were interpreted in terms of Lamarckism, natural selection or some other way. So, Göte Turesson and some of his followers expressed ideas later referred to as "bad Darwinism" (Komarov 1944) or typological thinking (Mayr 1982). They differed from "good Darwinism" in downplaying the role of the mutation pressure. The differentiation of a species was considered as a manifestation of previously existing forms rather than a formation of something new.

Thus, numerous, and quite different evolutionary concepts existed in the 1900–1930s. None of them seemed to dominate. Correspondingly, "crisis" was interpreted differently at that time, namely, as a coexistence of several contradictory evolutionary concepts, each of which faced significant objections (Rensch 1929). Some of these concepts were associated with orthogenesis. They developed quite actively even if they were somewhat overshadowed by others. Several evolutionists presented new concepts of directed evolution, while numerous other scientists expressed support of orthogenetic ideas. One might say that in the 1910–1920s orthogenesis was at the peak of popularity.

One of the largest contributions was made by a Russian scientist **Leo S. Berg** (1876–1950), the author of the concept of nomogenesis. Following his childhood interests, Berg started as an ichthyologist but early in his scientific career showed himself as a polymath. Having spent several years as an observer of fisheries at the

Aral Sea, Berg wrote a solid treatise, which, in addition to ichthyology, also dealt with geography and geology, and contained a general description of the Aral Sea and the surrounding territory. In 1909 Berg presented the book as a master's thesis but the committee considered it to be such a valuable contribution that it awarded Berg the doctoral degree (Krupennikov 1976; Isachenkov and Kvasov 1988).

As the scope of his studies expanded, Berg addressed the problems of evolutionary biology. His major works on evolution were written during the period of revolutions and the Civil War in Russia. In 1922, when scientific institution started to function more or less normally, he published his book *Nomogenesis* (Berg 1922a) and an essay *Theories of evolution* (Berg 1922b cited in Berg 1977) presenting a historical review of evolutionary views and a sketch of nomogenesis. Berg called his concept nomogenesis, meaning evolution determined by law, and referred to Darwinism as tychogenesis, that is, evolution directed by chance events. In 1926 *Nomogenesis* was translated into English (Berg 1926). In the 1960–70s its new editions were published in Great Britain and the USA. However, Berg stopped to write on evolution soon after the publication of *Nomogenesis* in Russian though he continued to work in geology, geography, history of sciences and biology (mainly, ichthyology).

Berg left an impressive scientific heritage. He is considered as the founder of Russian ichthyology (studies on Russian fish fauna still begin with references to his works). He is equally well-known as a geographer (it was as a geographer that he was elected member of the Academy of Sciences). Berg was a also person of encyclopaedic knowledge and diverse research interests ranging from climate change, principles of landscape division, glaciers, the formation of loess and the origin of land plants to the origin of Aleuts.

The publication of *Nomogenesis* was a remarkable event in the Russian biology. It received an enthusiastic response and strong criticism simultaneously. The laws of evolution attracted great attention, and meetings devoted to directed evolution were organised. Nomogenesis was especially well-met by Russian botanists. A new article about nomogenesis written by Berg was even published in the proceedings of applied botany (Berg 1925) though it was in stark contrast with other papers. Some Darwinians even complained about the "dominance" of nomogenesis in Russian biology (Kozo-Polyansky 1923) but these complaints were hardly justified. In general, a critical attitude to nomogenesis seemed to prevail from the start and progressed rapidly (Ivantsov 1923; Kozo-Polyansky 1923; Duchinsky 1925), while several initial sympathisers (Nikolai I. Vavilov, Theodosius G. Dobzhansky) later renounced their support.

A collection of papers criticising nomogenesis with the use of ideological arguments was published in 1928 (Kozo-Polyansky 1928). Nomogenesis was labelled a non-materialistic doctrine. In such a situation Berg could not possibly continue his evolutionary studies. In the 1930s he was involved in the publication of Darwin's books in Russian but dealt only with zoological issues and hardly expressed any of his own evolutionary views (Berg 1936). The collection of Berg's papers published in the 1950s (Berg, 1956–1962) did not include any of his evolutionary works either. (This was partly compensated for by a reprint of *Nomogenesis* in 1977.)

However, nomogenesis has not been forgotten. It repeatedly emerged in the works of Russian scientists. Many of them, including some historians of biology, believed that Berg was the first author of the concept of directed evolution or that, even if there was someone else in Europe, Berg was still the most important author in this field. On the contrary, European historians are surprised to find that someone has even suggested this concept in Russia. Berg himself disposed of the sensitive topic of priority in a single elegant gesture. He withdrew from the competition for priority from the start, declaring his readiness to concede it to anyone who would claim it. He referred to numerous studies where his ideas might have been anticipated, noting, for instance, that observations and experiments of Nikolai I. Vavilov were a better illustration of nomogenesis than anything that he, Berg, could offer (Berg 1977, p. 224). At the same time, his manner of presentation indicated that nomogenesis was a novel evolutionary concept and the new name he coined added to the impression.

A common theme of nomogenesis is an opposition to Darwinism. (But not to Darwin personally. Berg immensely admired Darwin as a naturalist and a man even though he completely disagreed with him in views on evolution.) Simplifying the work of the future researchers of his scientific heritage, Berg gave the main points of his concept in ten succinct theses:

(1.) Organisms evolved from thousands of primary forms.
(2.) Further evolution mostly proceeded by convergence (partly by divergence) and
(3.) was determined by laws
(4.) affecting numerous individuals over vast territories
(5.) in discrete steps.
(6.) The number of hereditary variations is limited, and they follow definite directions.
(7.) Struggle for existence and natural selection are not factors of evolutionary progress but conservative forces protecting the norm.
(8.) Owing to their mutational origin, species are distinctly separated from each other.
(9.) Much of evolution is revealing of a pre-existing potential.
(10.) Extinction is caused both by internal and external factors. (Berg 1977, p. 311).

Besides these ten theses, nomogenesis contains the idea of the direct action of the environment. According to Berg, some differences between species and intraspecific categories result from the action of the "geographical landscape". This explains adaptation, which is, however, not the major direction of the evolution.

Berg generalised an enormous bulk of data from genetics, morphology, taxonomy and palaeontology. He could discuss equally well the parallelism of flowers and the reproductive organs of gnetaceous plants, palaeontological studies by Edward Cope, experiments by Hugo de Vries, anthropological studies by Franz Boas etc. He touched upon most major achievements of contemporary biology and cited both modern and ancient scientists and philosophers. The bibliography of nomogenesis contains hundreds of sources. It is interesting that Berg, aiming to present a balanced survey of evolution, seldom used his own data on fishes, though they could be used to prove his point. Such an approach confused greatly some of his critics, who were forced to acknowledge Berg's erudition.

Berg proceeded from the assumption that any chance event is a manifestation of a regular process. For example, a person can accidentally send a letter without writing the address. However, a chance event like this occurs with a certain frequency. In 1908, 1.2 billion of letters were sent in Russia, with 27 letters per million being unaddressed. In 1910, 1.5 billion of letters were sent, with the same 27 per million were without addresses. Turning to biological phenomena, Berg disagreed that the emergence and distribution of organic characters were governed by chance. His objections were mainly based on similarities between groups of organisms inexplicable by direct relationship. He collected a huge number of such instances: electrical organs in different groups of fishes, hornless cattle originated independently on different continents, very similar shells of extinct cephalopods from different branches etc. From Berg's viewpoint, these data indicated that evolution proceeded mainly by convergence. If so, adaptation was merely an insignificant by-product of this process. Berg criticised many adaptationistic explanations. For example, the Darwinian interpretation of the origin of flying fishes means that they have adapted to flight, i. e. because of selection pressure the fins of the ancestors of these fishes became elongated, the fishes became capable of jumping, and now we can observe an original construction: the flying fish. Berg rejected this explanation, asserting that fins could be elongated irrespectively of any selection pressure and requirements of the environment. There are fishes with normal fins which jump from the water and fishes with elongated fins which never do so, living near the bottom. Berg argued that long fins are one of the possible kinds of variation, which is typical of some fishes and occurs as an anomaly in some others. He himself noticed such an anomaly in the roach (*Rutilus rutilus*) in the Aral sea (Fig. 3.1). Anomalies originate spontaneously, irrespectively of the environment, though they seem like adaptations under some conditions.

Berg collected abundant material aiming to prove the existence of spontaneous evolution in a definite direction but could say almost nothing about the mechanism of evolutionary transformation. He just suggested that definite directions in evolution were caused by "stereochemical properties of proteins". Berg never hinted at any mystical or non-material forces driving evolution but critics still accused him of religious obscurantism. In a caricature published in the newspaper of the Leningrad University in the 1930s, Berg was shown in a Russian national dress holding in his hands *Nomogenesis* and *Im Westen Nichts Neues* (*All Quiet on the Western Front*, a novel by Erich Maria Remarque about the First World War). Marxist ideologists sensed that nomogenesis implied pacifism and other things incompatible with the Soviet ideology. These accusations were partly due to the fact that Berg was a *tolstovets*, a follower of the writer Leo N. Tolstoy (1828–1910), who preached "non-resistance to evil by force", mutual love, equality of rights, vegetarianism, simplicity of family life, rejection of luxury etc. These views came into prominence in the 1880s, serving as a basis for a peculiar doctrine reminiscent of both Buddhism and Christianity. *Tolstovtsy*, considered as a heretical sect by the Orthodox Church, were

Fig. 3.1 An anomalous roach found by Berg and a flying fish (Berg 1977)

persecuted by the Tsarist administration and later by the communist regime. Berg indeed tried to follow Tolstoy's principles. He was a vegetarian and an ascetic. He started to rear his children in an unspoilt environment of a remote Russian village but the Revolution destroyed this idyll (Berg 2003).

Berg's opposition to Darwinism was not only scientific but also moral. He foresaw the establishment of totalitarianism in Russia and thought that its ideology corresponded to the idea of progressive role of struggle for survival in the evolution. Berg wanted to resist it as best he could. Nevertheless, the attacks of Soviet ideologists were ill-founded because Berg kept his philosophical views in his family and did not let any of them seep into his scientific work. One cannot guess Berg's life style from his writings. He never mentioned Tolstoy's principles in his evolutionary treatises and, unlike Tolstoy, did not attract attention by wearing the Russian national dress. Berg always stressed that nomogenesis was a purely scientific concept. He pointed out that Darwinism could not be accepted as a perennial dogma and repeatedly warned about the dangers of dogmatic thinking. He chose Huxley's statement "Science commits suicide when it adopts a creed" as an epigraph to the Russian edition of *Nomogenesis*.

Berg's appeal to reason hardly impressed the ideologists, who instigated his persecution in the late 1920s. Isaac Present (1902–1969), later a fellow-champion of Lysenko, was especially active. Fortunately, Berg, though persecuted, was never repressed, probably because he never replied to the accusations. In the 1930, when the persecution reached a climax, Berg left the University of his own volition. The faculty of geography began to collapse without him, and the university leaders asked Berg to come back, which he did it in 1936. By that time passions had calmed down a little, and Berg tried not to provoke them anymore. He continued his studies in geology, geography, biology (mainly ichthyology) and the history of a science, reaching the highest academic positions. In 1940 he was elected President of the Russian Geographical Society, and in 1946, the member of the USSR Academy of Sciences.

A Russian palaeontologist **Dmitry N. Sobolev** (1872–1949) elaborated similar views on evolution in parallel to Berg or even earlier. Before the First World War, he wrote a book on goniatites (an extinct cephalopod group), which contained several orthogenetic ideas. Sobolev submitted it as a doctoral thesis in the Kiev University but it was turned down. It was rumoured that the board was displeased with "inappropriate" citations from the poems of Lucretius. Sobolev refused to believe that such a minor reason could affect the judgment of competent scientists and explained their displeasure by the "heretical character" of his thesis presenting an uncommon view of evolution.

The "heretical character" was expressed not so much in the citations from Lucretius but rather in references to Theodor Eimer, Gustav Steinmann, Carl von Nägeli, Edgar Dacqué and Jean-Baptiste Lamarck, the proponents of directed evolution. Sobolev hardly ever mentioned Darwin's theory considering it unrelated to the explanation of evolutionary novelties. Sobolev regarded the popularity of the theory of natural selection in this field as a misunderstanding: "Darwin's followers proclaimed the omnipotence of natural selection in the origin of the new organic forms. However it sounds rather strange because even Darwin himself did not aspire to explain the origin of novelties, ascribing it to chance. He just explained the advantage of survival of those forms which have some characters suitable for the struggle for existence" (Sobolev 1914, p. 105). Discussing the origin of new characters, Sobolev also considered mutations and combinations, which were also opposed to Darwinism at that time.

Besides the neglect of Darwin's writings, the "heretical character" became apparent in the Sobolev's approach to taxonomy. Confronted with a great variety of cephalopod shells, he looked for formulas and laws to overcome the chaos. Sobolev considered living organisms as a kind of "isomorphic mixtures" of chemical substances, which should be decomposed down to their components to clarify their classification.

Sobolev's book was extremely original, which, rather than anything else, might explain the displeasure of his senior colleagues. One cannot help admiring Sobolev's principles: he did not change his viewpoint or compromise but continued to develop his concept. Eventually his merits were appreciated. Despite the failure of the doctoral thesis, Sobolev became head of the geology department of the Kharkov

University, occupying this position until his death in 1949. From 1922 up to 1933 he was also director of the Research Institute of Geology, and in 1934–1935, the dean of the Faculty of Geology and Geography. In 1934 he was finally awarded the doctoral degree for life-time contributions to geology (Yakovlev 1954–1955; Lapkin and Remizov 1972).

The book *Principles of historical biogenetics* (Sobolev 1924) contained ideas close to nomogenesis. Other books followed (Sobolev 1926, 1927a, 1927b, 1928). Their first versions were mostly written during the times of war and revolutions but Sobolev did not consider them out-of-date and made no essential corrections to these later editions. In the years that followed, Sobolev's non-Darwinian heresies remained separated from his scientific career (which reminds us of the case of Berg). He was an acknowledged expert in historical geology and palaeontology (although in these fields he also expressed many "disputable" ideas) and his "historical biogenetics" passed largely unnoticed.

Unlike Berg, Sobolev continued theoretical research on evolution though not on the "Darwinism versus orthogenesis" topic. He became interested in the global evolutionary processes involving the entire biosphere. Sobolev paid little attention to natural selection and variation but the core of his early concepts—evolution based on laws—hardly changed.

According to Sobolev, the main question concerning evolution was: "Is it an accumulation and selection of random variations occurring in all possible directions, which in themselves do not predetermine a defined course of the phylogenetic trend directed by selection only or is this elementary variability ordered and subordinated to certain laws predetermining the direction of evolution irrespectively of selection?" (Sobolev 1924, p. 24). Sobolev resolutely supported the second viewpoint. The claim of evolution on the basis of laws was his central point. His major book about "historical biogenetics" presented "the laws of biogenesis". Sobolev's biogenesis was roughly the same as evolution but Sobolev preferred to use the term "evolution" in its original sense of the process of development of something preexisting. Indeed, at least before the turn of the twentieth century the term "evolution" was used either as a synonym of preformism or to denote individual development. If so, "Darwin reaped a fine crop where he had not sown, he became famous as a founder of evolutionism without being an evolutionist" (Sobolev 1924, p. 32). His concept got the name of evolutionary doctrine by sheer misunderstanding.

This is not the only example of terminological confusion in Sobolev's writings. For some reason he liked to meddle with terminology, replacing "biology" with "biontology", "palaeontology" with "historical biontology" etc. "Historical biogenetics" in the title of his book means the study of "biogenesis", that is, "normal successive change of genetically related living forms" (Sobolev 1924, p. 3). Sobolev also coined terms for taxonomic categories of cephalopods, including numerous amusing "military" terms. He described "regiments" of ammonites splitting into "bands", "cohorts", "phalanxes" etc. creating an impressive image of a full-scale military offensive by various quaint creatures on the modern world. Entire columns of them perish, as in a world war, and new ones come instead. Some organisms survive by chance, but the largest and maybe the best part of them disappears forever.

Such a style was unlikely to make the author popular in the scientific community. Still, one can find in Sobolev's writing valuable ideas concerning the ordering of data on variety. He tried to reveal in living organisms something similar to the periodic system of chemical elements or the homologous series of organic substances. He ordered his "divisions", "maniples", "phalanxes" etc. in parallel series. In 1927, Sobolev summarized his studies on goniatites presenting a "reticular gradational-combinational system". He tried to combine the genealogical and the typological approach inventing systems reflecting typological affinities but containing some "historical aspects". For this purpose, he composed numerous tables of combinations of characters: vertical columns contained "steady characters" which originated by saltation, while horizontal lines contained "gradational characters" which originated in the process of evolution in definite directions. He noted that "Periodicity is not alien to my system since during transition from one horizontal series to another similar transformations are observed" (Sobolev 1927b, p. 172). Unfortunately, throughout his "periodical system" he persistently used his unhappy "militarised" terms.

Sobolev remarked that his concept corresponded to nomogenesis in general but was different in some details. The most essential difference concerned the problem of irreversibility.

Berg believed that "the possibility of reversible processes and, thus, repeatability is inherent in the living matter. The world of dead matter is subordinate to the law of dissipation of energy and is thus ultimately non-repeatable, while the world of living matter, being continuously in the process of struggle against dissipation of energy is potentially repeatable" (Berg 1919, p. 293). However, according to Berg, the potential repeatability is hardly ever realised, and the evolution in general tends to be irreversible. Berg believed that the bulk of experience accumulated by palaeontology was a brilliant illustration of Dollo's law. On the contrary, Sobolev believed that the reversible movement was an integral part of evolution just as well as the movement forward. It is known that a specialised organism originates from a less specialised ancestor. However, all organisms, including relatively "non-specialised" ones, have some features of specialisation. Sobolev agreed with Cuvier in that "nature could not be populated by schemes". This begs the question: what is the ancestor of a specialised group? It must be non-specialised but in the same time it must not be a mere "scheme". For Sobolev, the paradox was easily explained: the ancestor of a specialised group is specialised too but it can return to an earlier stage of its evolution and then start to evolve in another direction. This process occurs so rapidly that it is impossible to find transitional forms in the fossil record.

In the last writings by Sobolev, another significant difference with nomogenesis became apparent. While Berg expected to find the solution to the problem of evolutionary mechanisms from studies of organic molecules, Sobolev looked to the highest levels of the living world. He came to the conclusion that it was impossible to explain the mechanisms of evolution within the framework of research of isolated organisms. Only the analysis of the evolution of the entire Earth could clarify this problem. Sobolev worked on this subject during the last years of his life, planning to write a special treatise about it but did not have the time to complete this work.

Sobolev considered evolution as a cyclic process, and discussed various stages in the life cycles of groups of organisms or faunistic complexes. He analysed not only the origin of phylogenetic trends but also their "senescence". He believed that there were few examples of extinction resulting from the struggle for existence and, moreover, that it was in general difficult to explain extinction. "One always has to admit the extinction of forms distributed worldwide without any apparent involvement of any extinction agents arriving from somewhere" (Ibid., p. 180). An idea of the senescence of taxa suggests itself. However, Sobolev believed that a mere idea of senescence, dating back to the works of J. Brocchi in the early nineteenth century, was too general and contributed little to the understanding of this phenomenon. The idea of Daniele Rosa that extinction was due to a decreased capacity to vary was slightly more definite. Yet there are examples of both long-living non-variable forms (the horseshoe crab) and, on the contrary, variation bursts preceding extinction (Climenia).

Sobolev preferred to explain extinction by overspecialisation or non-adaptive trends in evolution. Analysing the "harmony" of the organisms' structure, he came to the conclusion that since organisms were complicated systems, their harmony could either increase or decrease. A "decrease of harmony" was an indication of senescence. Sobolev illustrated this idea with such organs as large horns of elks, fangs of sabre toothed tigers and mammoths' tusks.

However, a decrease of harmony did not seem to Sobolev a fully satisfactory explanation. After "historical biogenetics" he published a special paper on extinction in which he discussed this phenomenon in another light (1928). The main causes of extinctions should be looked for in "diastrophes", periodic movements of the Earth's crust resulting in perturbations of its surface. Sobolev was sceptical about hypotheses explaining extinctions by environmental changes. He believed that they were applicable only to a few cases and not to extinctions of widely distributed large animal groups. Sobolev tended to think that geological changes concerning the whole Earth violated harmony, which influenced the entire living world. The "violation of harmony" resulted in extinctions of some groups (mainly those that had already become "disharmonious") and the others occupied the free space.

Sobolev's "biogenetics" was less vulnerable to criticism than Berg's nomogenesis, and the attempts of Soviet critics to discredit it were less successful. This was partly because Sobolev's biography did not contain such a spicy detail as *tolstovstvo*; on the contrary, Sobolev resolutely abandoned the religious family tradition (he was born in the family of a priest). Similarly to Berg, Sobolev did not allow himself any hints on nonmaterial forces in his scientific research. His writings, which were full of tables, formulas and laws, looked purely scientific. For these reasons, critics tended to consider biogenetics as a mere eccentricity. It seems, however, that Sobolev did have some "non-scientific" ideas driving his research but they were not Christian, as in the case of Berg, but pre-Christian. Sobolev had a profound knowledge of ancient languages and was greatly influenced by ancient philosophy. Perceiving Darwin's theory in context of many centuries of cultural heritage, he considered it as a small episode in the history of science, which was not so significant as propagandists claimed. His attitude to Darwinism was condescending rather than critical.

Several other authors in various countries suggested ideas of directed evolution at the same time as Berg and Sobolev. An American scientist **Henri Fairfield Osborn** (1857–1935) actively developed these ideas and presented them as a special concept. Osborn was mostly known in the scientific community as a palaeontologist. In some respect, he can be considered as a new founder of American palaeontology, which had been dominated during several decades by the "war of the two giants", Edward Cope and **Otheniel Marsh** (1831–1899), with some involvement of the third "giant", **Joseph Leidy** (1823–1891). Marsh and Cope were rich and could conduct extensive palaeontological studies at their own expense. Both hired specialists to work for them and published actively. At the same time, they tried to ignore each other's studies, sometimes presenting parallel descriptions of the same data. Leidy, while also being a competent palaeontologist, did not have the same financial means, and his data were sometimes described anew by his rich colleagues.

This meant that Osborn had received a large scientific "inheritance", which was, however, in need of a serious revision. He carried it out successfully, adding his own data. At the same time, he was engaged in other studies. Palaeontological papers make up about 1/3 of his total list of published works, the others being devoted to geology, zoology, embryology, neurology, psychology, anthropology, history of science, museums etc. (Borisyak 1936; Serebryakov 1936). Using this abundant material, Osborn repeatedly addressed the problem of evolution. Strictly speaking, he was the author of several evolutionary concepts because his views changed several times in the course of his life. His attitude to Darwinism was ambiguous. He had a great respect of Darwin and considered Wladimir O. Kowalevsky as his teacher in palaeontology (Kowalevsky was a vocal advocate of Darwinism and, probably, the first scientist to interpret fossil data in terms of natural selection theory—Popov 2016). Osborn wrote positively about natural selection theory in some publications but at least one of his concepts, aristogenesis, was an alternative to it (Osborn 1934).

In this concept, Osborn considered the origin of new organs as the main problem of evolutionary biology. Though many aspects of this phenomenon remained obscure, Osborn pointed out an aspect that he called aristogenesis (*aristos* means "the best of its kind"). Aristogenes, according to Osborn, were new adaptive units evolving slowly for the use in a remote future; they were not useful for an organism or a species but served as a basis for the evolution of organs important for the descendants of their owner. The evolution of aristogenes—that is, aristogenesis—was exemplified by horns of titanotherians, by teeth of proboscideans etc. The first example was especially demonstrative: the first known titanotherians had small prominences on the heads, which later evolved into small horns, and later still, into large horns of complex shapes. Apparently, these organs were useful for their owners only at the last stages of evolution, when they could be used for defence or for fighting each other (Fig. 3.2). The teeth of proboscideans illustrate, though less vividly, the same situation: at first there were small additional prominences hidden

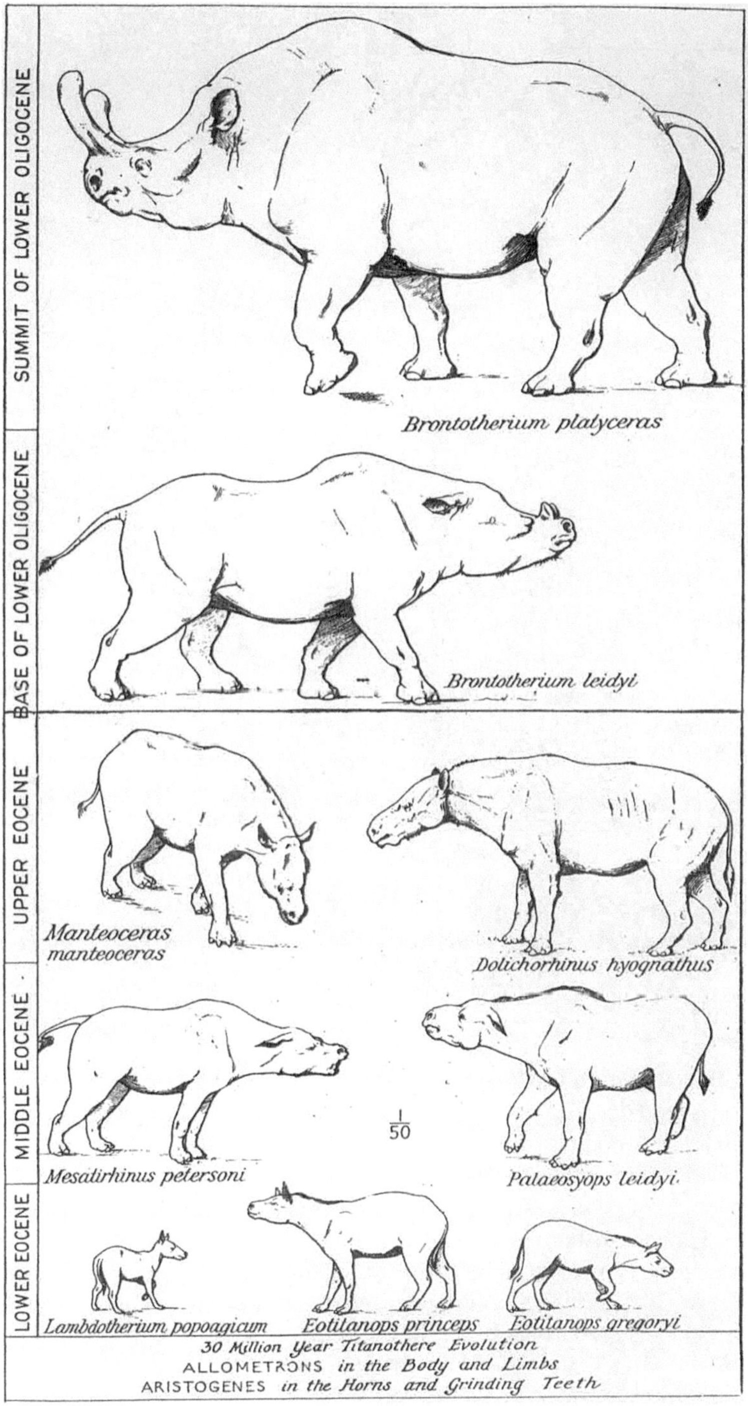

Fig. 3.2 Illustrations of aristogenesis: (a) evolution of horns in titanotherians; (b) evolution of teeth in proboscideans

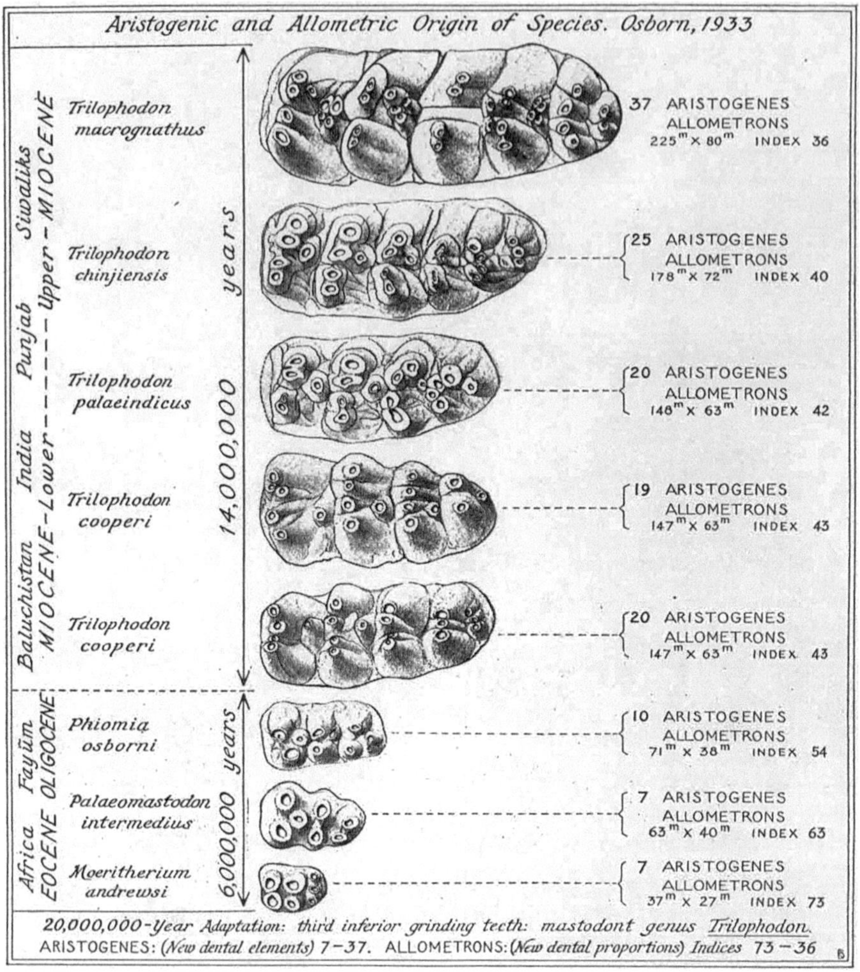

Fig. 3.2 (continued)

among the large ones used for mastication; later small prominences evolved into large ones and became suitable for mastication too. Discussing aristogenesis, Osborn used the notions of "potentiality" and "predisposition", indicating the predetermined nature of the process, and wrote approvingly about orthogenesis.

In addition of predisposition, aristogenesis was thought to have a "creative" and gradual character. Osborn believed that phylogenetic trends of various orders, families, genera and species could be characterised by certain aristogenes. He wondered what caused their origin from the germ plasma but the question remained unanswered, being as mysterious as the cause of the evolution of a rudimentary horn on the cranium of the titanotheres. "We know how these aristogenes arise, but we do not know why they arise, *i.e.*, their causes" (Osborn 1934, p. 210). "What evokes an aristogene from the gene-plasm is as mysterious to Us as what evokes a horn rudiment in the skull of the titanothere. We remain in the present part of this address purely on observational and inductive grounds and simply make statements of fact or of principle without offering any explanation." (Osborn 1934, p. 227).

At the same time, Osborn made some suggestions in this respect. He thought that aristogenes could not originate spontaneously but were triggered by some external influence. They might be the consequences of new conditions of environment or nutrition but whatever the case, the natural selection theory explained nothing in this respect.

Osborn offered some other arguments against Darwinism. For example, he pointed out that proboscideans did not encounter the problem of competition. They were the most powerful and marvellous animals on land, which nobody could compete with. (Osborn seems to have had a particular liking for elephants, mastodons and other proboscideans. He always wrote very warmly about them.) Nevertheless, they evolved, and did so very rapidly, even though they reproduced slowly. Darwin's theory envisages the opposite conditions for rapid evolution: fast reproduction, strong competition increasing selection pressure. (To note, some rapidly reproducing unicellular organisms amaze Darwinists by their constancy over hundreds of millions years—see Schmalhausen 1946).

Osborn interpreted the role of environment in evolution in approximately the same manner as it is usually done nowadays. He was sceptical about speculations concerning the inheritance of acquired characters. However, he thought it was impossible to deny the direct impact of the environment on some changes of organisms. This means that modification is either an evolutionary event or it can be replaced by a similar mutation in the process of natural selection. This view entered the history of biology as the "Baldwin's effect" named after James Mark Baldwin but Osborn suggested it independently (so did, incidentally, C. Lloyd Morgan).

Osborn did not reject evolution through adaptation to the environment completely. He even carried out an analysis of adaptation principles revealing its numerous kinds such as degeneration, compensation, development, compensation, economy, change of proportions, coadaptation, acceleration, retardation, self-adaptation, sports. He also suggested the "principle of adaptive radiation": dispersion of a group in various directions by adaptation to different environments (water bodies, lowlands, forests, grasslands etc.). This principle was often cited approvingly by zoologists and evolutionists.

To sum up, Osborn's views on evolution were rather close to the modern synthetic theory of evolution: he rejected the elements of Neo-Lamarckism of the nineteenth century and accepted natural selection and adaptation. It would seem that these studies could have led Osborn to modern Darwinism based on the genetics of populations but this was not the case. Osborn was critical of a too extensive use of genetics in evolutionary biology. He thought that genetics was about heredity, not evolution. According to Osborn, "palaeontology is evolution". Moreover, he believed that the adaptation principle could not be used to explain major evolutionary events. Adaptation concerns the change of already existing organs, not the origin of new ones. Characters changing in the process of adaptation were called by

Osborn “alloiometrons” and considered as more or less temporary reactions of evolving organisms to environmental changes. While they were not insignificant at a global scale, they did not explain evolution as a whole (Osborn 1934).

In parallel, another concept of directed evolution emerged in Italy. An Italian zoologist **Daniele de Rosa** (1857–1944) elaborated a “new theory of evolution and geographical distribution”—ologenesis (de Rosa 1931). Rosa’s own zoological studies dealt with taxonomy and morphology of Oligochaeta, but this was hardly reflected in his treatises on evolution as he presented generalisations concerning numerous taxa. The first version of his concept was published in 1909; then Rosa made several additions, translated his work into French and published the final version in 1931. *Olos* means *whole*, and ologenesis can be interpreted as “global development” or “global evolution”. Its main points are as follows.

1. The evolution of idioplasma of species, which transforms into phylogenetic evolution of organisms, has, similarly to all phenomena of the living world, internal and external factors but is not determined by the changes of the latter; it occurs even if external factors do not change.
2. The direction of evolution does not in the least depend on the variety of external factors.
3. In spite of this, phylogenetic evolution is not endlessly straight; on the contrary, it branches (dichotomically) because of differentiated divisions occurring in germinal cells as a consequence of the condition reached by idioplasma in the course of time.

As we see, Rosa used the notion of idioplasma, which had been proposed by Carl Nägeli. He considered Nägeli as his predecessor and named Kölliker and, partly, Lamarck as other forerunners of his theory. An earlier source of ologenesis was preformism. Rosa claimed that each species is predetermined in a previous one just as an individual is predetermined in the ovum (de Rosa 1931, p. VI). As for the term “orthogenesis”, Rosa used it in Eimer’s sense (to remind, Eimer considered the action of environment as a driving force of evolution). Rosa considered his concept among those that explained evolution by internal processes and criticised all other evolutionary theories. From his viewpoint, Darwin’s theory concerns changes within the limits of species or, at best, genera. Natural selection is similar to artificial one because it cannot create new groups of high taxonomic level, i.e. cannot explain phylogenetic changes studied by palaeontologists. Lamarckism (in its best known version, i. e. direct action of environment on evolution, exercises) is a mere hypothesis, yet unproven. Mutationism is also not a good concept, because nobody knows what “species” of de Vries are; they might well be no species at all. “I will not speak of various doubts that have been raised concerning the value of “species” which de Vries saw originating by ‘mutation’”; “We are still waiting for the proof that these mutations really give birth to new species”, Rosa wrote acrimoniously (de Rosa 1931, p. 34, p. 49). Apparently, he thought that the entire mutation theory was a nuisance.

Rosa compared the lifecycle of species to that of individuals. He believed that species were born, developed, reproduced, aged and finally died naturally. The aging of species was associated with the loss of variability. In the final stage of evolution, the variability of a species reaches a zero point, it becomes unable to respond to environmental the changes and becomes extinct. This "law" was hardly supported by evidence, however.

According to Rosa, there were even more similarities between the lifecycle of a species and that of a cell. He considered this idea as his most important contribution to evolutionary biology (de Rosa 1931, p. VII). Cells are known to divide inevitably and spontaneously when they reach a certain age due to internal processes. A species does the same. For this reason, evolution can be illustrated by dichotomic schemes. Rosa's schemes of this kind are one of the first prototypes of the now-fashionable cladograms.

According to Rosa, a species splits into two over its entire distribution area. This means that evolution is "ologenetic" (global): it involves all individuals of a species, not only selected ones. Ologenesis was thus suitable for the explanation of biogeographic data on. In this field, too, this concept aimed to substitute traditional views completely. Orthodox biogeography was based, according to Rosa (1931), on the "law of migration"[1]: new species originated from geographic races, formed as a result of dispersal and evolving under various conditions. Ologenesis stated the opposite. Rosa believed that differences in conditions could only determine the formation of "polymorphism" or "simple races". In this connection Rosa discussed the dilemma "polygenism vs. monogenism". (These terms are close to the terms polyphyly and monophyly but in this case they are used in a geographic sense: monogenism implies an origin of species in one point and polygenism, over a large area.) Rosa rejected monogenism resolutely, offering the following arguments:

1. Locally distributed modern species and groups had a large distribution area in the past.
2. This large distribution area (larger than that of modern local forms) did not originate from a smaller one or from a centre of origin.

For example, at that time the fossils of ancient placental mammals were known from USA, Europe and Patagonia. What should we consider as the centre of the origin of placental mammals? asked Rosa, implying that nobody could be so irresponsible as to point to a small area where the first placental evolved. Moreover, it is difficult to identify the centre of origin of several modern groups. Tapirs mainly occur in Central and South America; there are three species of tapirs there, and their distribution area is quite large. It is logical to presume that these mammals originated somewhere in America. However, the closest relative of American tapirs inhabits a small area in the South-Eastern Asia. Rosa suggested a solution: the ancestors of tapirs occurred over a large area from Asia to South America, and at some moment they produced similar new species everywhere. The same is true of alligators: they mostly occur at the Atlantic side of America but an east-Chinese alligator violates several important zoogeographic principles.

[1] "Law of migration" (Migrationgesetz) was proposed by Moritz Friedrich Wagner (1813–1887), the author of the evolutionary concept of isolation (Wagner 1868).

Soon after Rosa published his work on ologenesis, this idea was picked up by a French researcher **Georges Montandon** (1879–1945). He accepted ologenesis unconditionally and wrote a large treatise on "human ologenesis", in which he analysed the origin of mankind and the genealogy of human races and civilizations. Not stopping there, he wrote a treatise on the ologenesis of cultures. He believed that "cultures" did not correspond exactly to "races" but evolve spontaneously by themselves. Montandon described "cultural cycles" much in the same way as Rosa described life cycles and branching of species. In this way, ologenesis was transformed into a global evolutionary and historical concept (Montandon 1928, 1934). However, Montandon later buried this trend of research. He collaborated with fascists during the Second World War, published papers supporting the Nazi regime and was finally killed by the French resistance.

Orthogenesis was rebranded yet again by an Austrian scientist **Hans Przibram** (1874–1944). His "morphogenetic concept" was called apogenesis meaning the loss of possibilities during differentiation. He noted that "this concept is nothing more than orthogenesis of Eimer" (Przibram 1929, p. 299). However, Przibram's apogenesis was in fact quite different from Eimer's views. Although apogenesis emerged after orthogenesis, it would seem that the contrary was the case. Out of all concepts of directed evolution, apogenesis seems the most old-fashioned. It is reminiscent of the old debates concerning preformism and epigenesis of the eighteenth century. Apogenesis was proposed as an alternative to both. Preformism implies that the ovum is complex, while epigenesis implies that it is simple. Przibram thought that the ovum was complex but also thought that its complexity dispersed gradually. At the same time, he admitted, similarly to epigeneticists, that the organisms increased in complexity during development, not forgetting that potential possibilities dispersed. It is sometimes difficult to distinguish the analysis of embryogenesis and that of evolution in such apogenetic speculations. Przibram was inclined to deny evolution in the modern sense. He rejected both the transformation of one species into another and the existence of genealogical trees. He believed that the major animal groups evolved in parallel rows, considering this process as analogous to the growth of crystals (Przibram 1906, 1929).

Allelogenesis, an idea suggested by a French zoologist **Alphonse Labbé** (1869-?), is somewhat similar to the concepts described above. Labbé did not try to generalise everything but aimed to explain adaptation and some aspects of the evolution of invertebrates (Labbé 1924). His evolutionary views were mainly based on the data on crustaceans and salinity changes. The effect of salinity on them can be so considerable that crustaceans of the same species living in different conditions were often described as different species or even genera. Labbé supposed that the major invertebrate groups could have formed in this way. The main point of Labbé's concept was that the changing pH may result in the emergence of reversible or irreversible isomorphs. "So, one and the same ovum, because of an increasing PH, if it does not die, may give a product identical to the parent or a totally different one. This is the definition of allelogenesis." (Labbé 1924, p. 298). "Allelogenesis" is hardly related with the genetic term "allele", although it also means "one and the other" way of variation. To the two explanations of adaptation mechanisms existing at that time, selection and inheritance of acquired characters, Labbé added the third one: allelogenesis, stating multiple patterns of species transformation. He emphasised that this transformation had certain limits and could be described as orthogenesis without any finalistic sense. Labbé did not say that the change of environment caused allelogenesis but considered the external influence as a trigger of development of morphological characters (Labbé 1929).

To sum up, at least six new orthogenetic concepts emerged in the 1900–1930s. Directed evolution irrespective of natural selection was their central point. In this, they were hardly different from the concepts of the previous period but were presented as new theories with numerous new terms. The reasons behind this desire to change the existing terminology are unclear. Earlier works on directed evolution were well known at that time and were often cited by the new advocates of orthogenesis.

References

Berg LS (1919) Variability of a phenomenon and the laws of nature (Izmenchivost' yavleniya i zakony prirody). Priroda 7–9:292–302

Berg LS (1925) Regularities in the formation of organic forms (Zakonomernosti v obrazovanii organicheskikh form). Trudy po prikladnoy botanike i selektsii 14(5):19–68

Berg LS (1926) Nomogenesis or evolution determined by law. Constable & Co. ltd, London (English edition)

Berg LS (1922ab) cited in Berg LS (1977) Works on the theory of evolution (Trudy po teorii evolyutsii). Nauka, Moscow

Berg LS (1936) Charles Darwin as a zoologist (Charl'z Darvin kak zoolog). In: Ch D (ed) Collected works, 2. Zoological works. AN SSSR, Moscow/Leningrad, pp 9–36

Berg RS (2003) Hotwind. Reminiscences of a geneticist (Sukhovey. Vospominaniya genetika). Moscow

Borisyak AA (1936) Henry Fairfield Osborn (Genri Ferfild Osborn). Priroda 1:130–133

Bowler PJ (1983) The eclipse of Darwinism. Johns Hopkins Univ Press, Baltimore

Bowler PJ (2003) Introduction (Predislovie). In: In the shadow of Darwinism. Alternative theories of evolution in the 20th century (V teni darvinizma. Al'ternativnye teorii evolyutsii v XX veke). Yasnyy den', St Petersburg, pp 5–6

Cuénot L (1932) The origin of animal species (La genèse des espèces animales). Libraire Felix Alcan, Paris

de Rosa D (1931) Ologenesis. A new theory of evolution and geographic distribution of living beings (L'Ologénèse. Nouvelle théorie de l'évolution et de la distribution géografique des êtres vivants). Libraire Félix Alcan, Paris

Duchinsky FF (1925) Is Darwinism tychogenesis? (Yavlyaetsya li darvinizm tikhogenezom?) Pod znamenem marksizma, 8–9:116–125

Guyénot E (1921) Mutations and monstrosities (Mutations et monstrosités). Revue scientifique 21:611–617

Guyénot E (1930) Variation and evolution (La variation et l'évolution). G. Doin et C-ie, Paris
Huxley JS (1944) Evolution. In: The modern synthesis, 3rd edn. George Allen and Unwin, London
Isachenkov VA, Kvasov DD (1988) L. S. Berg. Prosveshchenie, Moscow
Ivantsov NA (1923) The factors of evolution (Faktory evolyutsii). Gosudarstvennoe izdatel'stvo, Moscow-Petrograd
Komarov VL (1944) The teaching about the species in plants (Uchenie o vide u rasteniy). AN SSSR, Moscow, Leningrad
Kozo-Polyansky BM (1923) The last word of anti-Darwinism. Presentation and critical analysis of the theory of nomogenesis, a new teaching about evolution of organic world (Poslednee slovo antidarvinizma. Izlozhenie i kriticheskiy razbor teorii nomogeneza, novogo ucheniya ob evolyutsii organicheskogo mira). Burevestnik, Krasnodar
Kozo-Polyansky BM (ed) (1928) The theory of nomogenesis. A new phase in the development of Russian anti-Darwinism. A collection of critical papers (Teoriya nomogeneza. Novaya faza v razvitii rossiyskogo antidarvinizma. Sbornik kriticheskikh statey). Izdatel'stvo Gosudarstvennogo timirhyazevskogo nauchno-issledovatel'skogo instituta, Moscow
Krupennikov IA (1976) L. S. Berg. Shtiintsa, Kishinev, pp 1876–1950
Labbé A (1924) A new concept of adaptation: allelogenesis (Une conception nouvelle de l'adaptation: l'allélogénese). Revue scientifique 10:295–301
Labbé A (1929) The transformist conflict (Le conflit transformiste). Libraire Félix Alcan, Paris
Lapkin IYu, Remizov NN (1972) Dmitriy Nikolaevich Sobolev (on occasion of the centennary of his birth) (Dmitriy Nikolaevich Sobolev [K stoletiyu so dnya rozhdeniya]). Izvestiya AN SSSR Seriya geologicheskaya 8:129–132
Mayr E (1982) The growth of biological thought. Harvard Univ Press, Cambridge
Montandon G (1928) Human ologenesis (ologenism) (L'ologenèse humaine [Ologénisme]). Libraire Félix Alcan, Paris
Montandon G (1934) Cultural ologenesis. A treatise on cyclo-cultural ethnology and systematical ergology (L'ologenèse culturelle. Traité d'ethnologie cyclo-culturelle et d'ergologie systematique). Payot, Paris
Nazarov VI (1974) Evolutionary theory in France after Darwin (Evolyutsionnaya teoriya vo Frantsii posle Darvina). Nauka, Moscow
Nazarov VI (1984) Finalism in the modern evolutionary teaching (Finalizm v sovremennom evolyutsionnom uchenii). Nauka, Moscow
Osborn HF (1934) Aristogenesis, the creative principle in the origin of species. Am Nat 68:193–235
Popov IYu (2016) Sources of the Darwinian palaeontology: Vladimir Kovalevsky and Louis Dollo. In: Biology and medicine in France and in Russia: historical crossroads (end of 18th—20th century) (Biologie et médecine en France et en Russie: Histoires croisées [fin XVIIIe-XXe siècle]). Hermann Éditeurs, Paris, pp 51–63
Przibram H (1906) Analogy with crystals for the developmental mechanics of organisms (Kristall-Analogien zur Entwicklungsmechanik der Organismen). Archiv für Entwicklungsmechanik der Organismen 22(1–2):207–287
Przibram H (1929) Apogenetical theory of the evolution of organisms (Théorie apogénétique de l'évolution des organisms). Revue générale des Sciences pures et appliquées 40:293–299
Rensch B (1929) The principle of geographical superraces and the problem of speciation (Das Prinzip geographischer Rassenkreise und das Problem der Artbildung). Verlag von Gebruder Borntraeger, Berlin
Schmalhausen II (1946) Factors of evolution (Faktory evolyutsii). AN SSSR, Moscow-Lenngrad
Serebryakov AE (1936) Henry Fairfield Osborn (Genri Ferfild Osborn). Priroda 1:134–138
Sobolev DN (1914) Sketches in phylogeny of the goniatites (Nabroski po filogenii goniatitov). Izvestiya Varshavskogo politekhnicheskogo instituta Imperatora Nikolaya II, 1:1–191
Sobolev DN (1924) The fundamentals of historical biogenetics (Nachala istoricheskoy biogenetiki). Gosudarstvennoe izdatel'stvo Ukrainy, Simferopol
Sobolev DN (1926) Earth and life. I. Geological cycles (Zemlya i zhizn'. I. Geologicheskie tsikly). Nauchno-populyarnaya biblioteka Ukr. Otd. Geol. Komiteta, Seriya 1, Kiev

Sobolev DN (1927a) Earth and life. II. Evolution and revolutions in the history of organic world (Zemlya i zhizn'. II. Evolyutsiya i revolyutsii v istorii organicheskogo mira). Nauchno-populyarnaya biblioteka Ukr. Otd. Geol. Komiteta, Seriya 1, Kiev

Sobolev DN (1927b) An experience of the construction of reticulated gradative-combinative system of goniotites (Opyt postroeniya retikulyarnoy gradativno-kombinativnoy sistemy goniatitov). Naukovi zapiski po biologii. Derzhavne vidavnitstvo Ukraini, pp 170–200

Sobolev DN (1928) Earth and life. III. On the causes of extinction of organisms (Zemlya i zhizn'. III. O prichinakh vymiraniya organizmov). Nauchno-populyarnaya biblioteka Ukr. Otd. Geol. Komiteta, Seriya 1, Kiev

Wagner MF (1868) Darwinian theory and the law of migration (Die Darwinishe Theorie and das Migrationsgesetz der Organismen). Duncker & Humblot, Leipzig

Yakovlev NP (1954–1955) Five years since the death of D. N. Sobolev (Pyat' let so dnya smerti D. N. Soboleva). Ezhegodnik Vsesoyuznogo paleontologicheskogo obshchestva 15:369–371

Zavadsky KM (1973) The development of evolutionary theory after Darwin (1859–1920s) (Razvitie evolyutsionnoy teorii posle Darvina [1859–1920-e gody]). Nauka, Leningrad

Chapter 4
Declarations in Favour of Orthogenesis in the 1900–1930s

Abstract The ideas of directed evolution were especially popular in palaeontology. They were also supported by idealistic morphologists, who described "typological affinities" of biological structures without analysing either their origin or their adaptive significance. This analysis revealed morphological regularities, which were used as arguments in favour of orthogenesis. Some taxonomists, zoologists, botanists, geneticists, biochemists also supported orthogenesis at that time, criticising the adaptationistic paradigm and revealing certain laws of variation and evolution. Some of their arguments still seem convincing.

Keywords Palaeontology · Idealistic morphology · Genetics · Orthogenesis · Zoology · Botany

In addition to the relatively new orthogenetic concepts, numerous declarations in favour of orthogenesis were made in the first three decades of the twentieth century. The ideas of directed evolution were especially popular in palaeontology. A German palaeontologist **Edgar Dacqué** (1878–1945) even noted that Darwinism had never had any true supporters among palaeontologists (Dacqué 1915). Similar sentiments were later expressed by Jacques Lessertisseur, who, reviewing the history of palaeontology and evolutionary biology, noted that although the evolutionary idea did settle in the palaeontology of the nineteenth century, "the palaeontologists had views, which differed significantly from those of Darwin" (Lessertisseur 1959, p. 158). Lessertisseur pointed out two groups of facts that were in conflict with Darwinism: 1) characters distinguishing species are non-adaptive and 2) the origin of new organs cannot be explained by selection of small variations. Both these groups of facts lie at the core of orthogenesis. If so, almost all palaeontologists of the early twentieth century can be considered as orthogeneticists.

An opinion about the initial unpopularity of Darwinism among palaeontologists sounds convincing. Until relatively recently (before the formation of modern Darwinism in the 1940s) palaeontological works often contained critics of the natural selection theory. However, there is also an opposite viewpoint: evolutionary biology (in the form of Darwinism) did triumph in palaeontology (Nazarov 1991, p. 62) and Lamarckians, such as Edward Cope, were in the minority until the twentieth

I. Popov, *Orthogenesis versus Darwinism*,
https://doi.org/10.1007/978-3-319-95144-7_4

century. Afterwards, however, (in 1900–1940s) palaeontology indeed became a bastion of anti-Darwinism (Davitashvili 1948). Who is right?

If we understand Darwinism as the writings of Darwin or ideas close to modern Darwinism, Dacqué and Lessertisseur were right. However, if we consider every statement in favour of natural selection as Darwinism, they were not quite right. Many palaeontologists approved of Darwin's theory but expressed views that were hardly compatible with Darwinism. So, **Louis Dollo** (1857–1931) believed that all living beings originated according to the "law" discovered by Darwin. At the same time, he shared the opinion about inevitable extinction of groups of organisms after the completion of a definite lifecycle. Moreover, he supported saltationism rather than gradualism. He claimed that "evolution is discontinuous, irreversible, restricted" (Dollo 1893, p. 165). **Melchior Neumayr** (1845–1890) also supported the natural selection theory but thought it absolutely improbable that variation should occur randomly in every possible direction (Neumayr 1919, p. 218). Variation follows definite directions, some of which are eliminated by natural selection while others progress. There was no contradiction between selection and palaeontology in this case. However, this was not "true" Darwinism because the opinion about unlimited or unexhausted variability was the basis of Darwin's theory. **Wladimir O. Kowalevsky** (1842–1883) was a vocal supporter of Darwinism and the first "Darwinian" palaeontologist but he also suggested "the law of inadaptive evolution" (the term was coined by his follower Leo Davitashvili—Davitashvili 1948). According to Kowalevsky, non-adaptive spontaneous evolution exists alongside with the adaptive one (Kowalevsky 1875). "Orthogenetic" paleontologists claimed the same.

A palaeontologist **George Gaylord Simpson** (1902–1984) was one of the "architects" of the modern evolutionary synthesis. He was, however, a kind of "blackleg", criticising and mocking his colleagues unrestrainedly. He remained aloof from his fellow "architects", and seemed to enjoy his aloofness, too. "Orthodox" Darwinists doubted that Simpson's views on evolution were quite right. The editors of the Russian translation of his book *Tempo and Mode of Evolution* noted that Simpson came to Darwinism "by the backdoor" of doubts and exceptions (Simpson 1948). Deviations from the traditional scheme can be easily found in his works. Like Neumayr, Simpson considered the view about variability occurring in every possible direction as patently absurd (Simpson 1944). He also remarked that it would be an exaggeration to say that the synthetic theory explained everything in evolution (Simpson 1950).

Orthogenesis was interpreted in a broadly different manner in palaeontological studies of that time. So, a German paleontologist **Gustav Steinmann** (1856–1929) believed that his support of Lamarck's ideas was "orthogenesis of the highest degree" (Steinmann 1908). Steinmann thought that he developed the idea of directed evolution more consistently than Cope and Eimer because he followed more strictly the principles of Lamarck's *Philosophy of Zoology*. Steinmann often cited Lamarck and dedicated his book to him. The essence of his "orthogenesis of the highest degree" was that every complex and specialised organism evolved only once during the history of the Earth and that the formative activity of nature was limited by narrow frames (Steinmann 1908). This means that the variability of species does not result in any significant evolutionary transformations. With time, an "organism", represented by variations, transforms into another. Similarly to Lamarck, Steinmann argued that there were no extinctions in evolution. If modern organisms differ from those that existed in the past, this means that they have undergone an orthogenetic transformation. Discussing presently non-existent forms, Steinmann preferred to use the word "disappearance" (*Verschwinden*) rather than "extinction" (*Aussterben*). From his viewpoint, a true extinction was possible only in exceptional cases and was associated with a negative external influence. For example, if a species was confined to a small island and if this island was destroyed by a volcanic eruption, the species would become truly extinct. Such events are rare and do not reflect evolution in general. Steinmann came to the conclusion that only humans could exterminate species. He pointed out the fact that many large mammals became extinct (or rather, "disappeared") during the Cainozoic era because they were valuable game to prehistoric hunters while less valuable mammals survived.

To prove that extinction was impossible, Steinmann tried to demonstrate the links between numerous animals that had "disappeared" and their modern relatives but without much success. To give just a couple of examples, he looked for the ancestors of vertebrates among trilobites and believed that dolphins descended from ichthyosaurs. His genealogical schemes look not like branching trees but rather like clumps of confervoid algae. In his opinion, an admission of polyphyletic origin of the living world would remove many unresolved problems of evolutionary biology.

Generally, Steinmann's phylogenetic reconstructions did not look convincing even to his contemporaries. Some authors, however, took him seriously. A Russian morphologist **Alexey N. Severtsov** noted that "the situation is not so easy as it seems" with Steinmann's works (Severtsov 1914, p. 71). Such an opinion might be due to the fact that Steinmann focused on numerous phenomena that were difficult to explain such as parallel evolution, decrease of variability, relations between mutations and variations, the absence of intermediate forms etc. Painstakingly comparing his hypothesis with the others, he finally managed to produce a modernised version of many of Lamarck's ideas. It is interesting that Steinmann scorned vitalism, and called evolutionary speculations vitalistic whenever he wanted to emphasise that they were weak and unconvincing (for instance, speculations about extinction or the absence of intermediate forms).

Karl Beurlen (1901–1985), another German palaeontologist, defended a different orthogenesis. Beurlen supported the concepts of Nägeli, Eimer, Döderlein and Rosa even though he challenged some of their aspects. In contrast to these authors, Beurlen rejected the direct impact of the environment. He focused his attention on explosive processes of the formation of types, which appear to have taken place in the past. In his opinion, such explosions give rise to both adapted and non-adapted forms. The latter die out very fast or are preserved as curiosities in small numbers. In his opinion, several such "failures", e.g. scaphopods, could be found among modern animals (Beurlen 1930). (It is unclear, though, in what way adapted animals are different from non-adapted ones.) According to Beurlen, an "explosion" was followed by an orthogenetic development of the new type either in progressive or in regressive direction. Beurlen lectured and published actively in the 1920s–30s and later, and discussed orthogenesis with his colleagues. Judging from these discussions, orthogenesis was quite popular among German palaeontologists (Levit and Olsson 2006; Rieppel 2012).

Another interpretation of orthogenesis was presented by a British palaeontologist **William Dickson Lang** (1878–1966). His studies mostly dealt with bryozoans of the Cretaceous period (Taylor 2002). According to Lang, spontaneous evolution —"orthogenesis"—meant an inevitable death of phylogenetic trends due to senescence. Lang cited Eimer approvingly but thought that the transformation of butterflies resulted from natural selection while a true orthogenesis was to be found elsewhere. He pointed that the thickening of walls between individuals in colonies occurred in the evolution of various bryozoan groups. A least 11 lineages evolved in this manner during the Cretaceous: the walls became so thick and the space between so small that the life of such organisms became unbearable and they died out soon after that. Storing calcium in the walls, these animals built graves for themselves. The most "promising" bryozoans had skeleton made of chitin. As soon as they got a skeleton made of calcium compounds, their destiny was predetermined. Lang discovered the same phenomenon in molluscs and brachiopods. Their evolution followed the same plan and resulted in a dead end. The origin of the plans remained an open question.

Lang's ideas about natural selection are greatly reminiscent of nomogenesis. Similarly to Berg, Lang came to the conclusion that natural selection was not a creative but a destructive force. Once bryozoans had "built a grave" for themselves, they were eliminated by natural selection. Natural selection plays a role in such cases only. It cannot create anything new and useful. These ideas were reflected in manuals on palaeontology and therefore persisted in the scientific community for a long time. However, he had no followers. Lang was often criticised by his close colleagues, who noted, in particular, that the accumulation of calcium by bryozoans could simply be the result of changes in the salinity in the seas where they lived.

The fate of Lang's evolutionary ideas was peculiarly similar to that of the ideas suggested by Berg and Sobolev. His view about evolution formed in the 1910s, and he published several papers in the 1920s. Then, in the 1930s, he stopped to mention evolution in his publications though continued to be active in research. Naturally, there was no such ideological control of theoretical studies in Great Britain as in Soviet Union but orthogenesis was still considered as something of a heresy. While Lang did not change his views on evolutionary mechanisms in the 1930s, he expressed them only in letters to his colleagues (Taylor 2002). Apparently he had difficulties with presenting them in scientific publications.

Evolution was usually represented in palaeontological works as series of regular stepwise transformations with numerous dead-ended branches. Fossil records are almost always discontinuous, with lack of intermediate forms between related groups and sudden origins of new structural types. Interpretations of this discontinuity may smack of creationism, and some palaeontological publications are on the verge of "good scientific manners" in this respect. So, Edgar Dacqué claimed that species and genera observed in fossil record were morphologically separated, closed types. Mutations, variations and other transformations were possible only within the types, while their origin could not be explained. "Types – so far as we can talk about them in an intelligible way – are potential species [Artpotenzen] realised in actual forms and underlying these forms. They are not simply abstractions from the concrete forms, but are ultimate genotypic realities and potentials [Potenzen] beyond the objective phenotypic realities" (Dacqué 1921, pp. 728–729 cited in Levit, Meister, Levit and Olsson 2006, pp. 288–289). Dacqué characterised the types as "realistic-unrealistic" forms. On the one hand, they are "genotypically" adjusted to the material world; on the other hand, types (or potencies) are emanations of an ideal world. Dacqué's types are analogous to the *eidoi* of Plato. They exist on the other side of the material world and sometimes surface in evolution. This process is not random but proceeds in definite directions. Discussing evolution, Dacqué stressed that its main trend was the one that led to the origin of man and used the term "orthogenesis" to describe this process.

The works of Dacqué reflect both orthogenesis and an "idealistic morphology", which was popular at that time. It was close to orthogenesis though it implied the analysis of morphology outside an evolutionary context. The term "idealistic" is frowned upon in science but in this case it did not mean anything non-materialistic. Idealistic morphology simply described "typological affinities" of biological structures without analysing either their origin or their adaptive significance (Blyakher 1962; 1976). This analysis revealed numerous morphological regularities, which could be used as arguments in favour of orthogenesis. Idealistic morphologists constructed schemes of regularities between types instead of traditional genealogical schemes. They often prided themselves on the "purity" of their morphology, unalloyed with any evolutionary speculations.

The principles of idealistic morphology stroke a chord with some taxonomic studies since the construction of phylogenetic schemes was problematic for several groups. Similar ideas were expressed independently of idealistic morphologists by Nikolay M. Gaydukov (1874–1928), who dealt with the taxonomy of algae. Initially he tried to base taxonomy on phylogenetics but this turned out to be impossible as there were no data to confirm phylogenetic schemes (Gaydukov 1925, 1926). Gaydukov divided algae into five types, the main criteria being the patterns of cell division and, correspondingly, those of cell structure. He showed that the same forms (branching, filamentous) could be achieved in various ways. Analogously, during production of technical items, a ring can be made of metal, wood etc.

Gaydukov began his studies of algae as early as in 1898. In the 1920s he generalised them and presented his views on taxonomy and evolution. He emphasised the role of convergences in evolution and put them at the basis of taxonomy of algae and higher plants. He cited Berg and Vavilov approvingly but preferred stay away from collision of Darwinism and orthogenesis. He strived for synthesis: "It is to be hoped that endless embittered disputes of Darwinians, Lamarckians etc. will finish one day, and we will have a broad biological evolutionary theory, which would be able to explain the phenomena by various and probably yet undiscovered factors" (Gaydukov 1926, p. 283). Gaydukov thought that the criticism of Berg by Darwinians was a misunderstanding since the two concepts were complementary. Darwin's theory considered the origin of changes as an accomplished fact while nomogenesis focused on their prerequisites. Gaydukov himself considered evolution as a result of numerous processes including competition, inheritance of acquired characters and spontaneous transformations. He was quite "Darwinian" in his interpretation of convergences: "the same physiological functions and the same ecologic conditions result in the same mechanically built structures irrespective of their onto- and phylogenetic affinity" (Gaydukov 1926, p. 270).

Several taxonomists and morphologists, which used traditional approaches in their studies, also expressed "orthogenetic sympathies". Brothers **Fritz Sarasin** (1859–1942) and **Paul Sarazin** (1856–1929) discovered the "orthogenesis" of molluscs (Indonesian snails) but noted that it was a descriptive rather than an explanatory term. Actually, they discovered a phenomenon that was later called "clines"—gradual transitions from one kind of variation to another. The Sarasin brothers described numerous series of transformations of snails, similar to those of butterflies in the works of Eimer. They concluded that neither species no variations existed in reality; there were just series, or chains of forms and some of their sections were called species or variations for convenience. These "chains of forms" (*Formenketten*) were considered as something strictly ordered, and natural selection was not even mentioned in this context (Sarasin and Sarasin 1899).

Bashford Dean (1867–1928), who studied chimeras (Chimaera) and other vertebrates, came to the conclusions reminiscent of aristogenesis. He noted that all evolutionary theories implied direct links between parents and offspring. In reality, however, various stages of development required various adaptations but the structure of the early embryonic stages was better adapted to the conditions experienced by the organism at the final stages of embryogenesis. In this way, characters are "foreseen". Studying embryonic capsules of chimeras, Dean found features that were neither adaptations nor provisional characters but indicated the needs of the organism (or species, or genus) in the remote future. The main conclusion was: "The capsule of *Chimæra* must stand, I conclude, as an instance of evolution in a determinate direction. Unhappily, however, it gives no data by which the cypher of orthogenesis can be simplified" (Dean 1904, p. 112).

A prominent German taxonomist, ornithologist and evolutionist, **Erwin Stresemann** (1889–1972), supported orthogenesis in the 1910s. Discussing similarities between tropical birds from different families, orioles (Oriolidae) and sunbirds (Meliphagidae), he claimed that the explanations of these similarities in terms of mimicry or natural selection were a fantasy. There was no selective factor in this case, and the similarity of colour was observed even if an area was populated by the representatives of only one family, i. e. the object of imitation was absent. Stresemann came to the conclusion that the similarity of these birds was "the result of an unrelated convergence" in response to "environmental stimuli" (Haffer et al. 2000, p. 222). Later, however, Stresemann changed his views on evolutionary mechanisms, embracing first mutationism and then the ideas of modern Darwinism (Stresemann 1975).

An Italian zoologist **Giuseppe Cei**, who studied mole rats, found that different species of these underground rodents demonstrated various stages of reduction of eyes. Cei considered these cases as orthogenesis or "phyletic tendencies". Should anyone described this material now, it would probably be in the light of adaptations, specialisations or even selection pressure. However, an orthogenetic interpretation of these data was apparently possible at that time (Cei 1946).

Orthogenesis was often mentioned in studies of various cave-dwelling animals (fishes, crustaceans, insects). Such studies have been referred to as an anomaly in the evolutionary biology (Bocking 1988). Similarly to any anomaly, the cave fauna was considered either as a curiosity not to be taken seriously or, on the contrary, as a window into the secrets of evolution. It first entered evolutionary biology in the times of Darwin. Agassiz used this material to criticise the idea of evolution, while Darwin supported his theory with it. Darwin thought that cave animals were relics of an ancient fauna which had survived under conditions of low competition. However, this was later disproved since some caves were shown to be recent.

In the late nineteenth century Alpheus Packard interpreted the results of his studies of caves in terms of Lamarckism (he introduced the term Neo-Lamarckism, independently of other authors, in the context of these studies). Packard's Neo-Lamarckism was especially close to that of Lamarck himself (Bocking 1988). The features of cave-dwelling animals such as the reduction of eyes, highly developed tactile organs and pale coloration were considered as a confirmation of the evolutionary significance of exercises or lack thereof. Later the material on cave fauna was used to support the idea of the direct action of environment on evolution. So, a French zoologist **Louis Fage** (1883–1964) demonstrated that cave spiders were the result of a direct action of underground conditions on spider metabolism (Fage 1931). However, in the twentieth century an orthogenetic interpretation of the cave fauna became more popular: cave-dwelling animals were considered as a display of evolutionary tendencies that could not be revealed under "normal" conditions. An American zoologist **Arthur Banta** (1877–1946) was the first to present this interpretation. He was strongly opposed to Neo-Lamarckism and considered preadaptation and predetermined variation as the mechanism of the origin of cave animals (Banta 1907). He rejected the idea that caves were populated in a random manner. According to his observations, various "normal" animals sometimes got into caves but could not settle there, that is, could not carry out exercises and adaptations. Cave animals originated from ancestors inclined to life in dark and wet places. At the same time, some species seem to be preadapted to caves but do not occur there. The formation of the cave fauna was, according to Banta, a strong case for orthogenesis.

A French scientist **René Jeannel** (1879–1965) suggested a compromise between orthogenesis and Lamarckism concerning cave animals. He was attracted to the idea of "somatic induction" believing that species transformed because of the organism's reaction to external conditions rather than to spontaneous changes in the ovum. Jeannel stated that modifications concerned not only somatic cells but germinal ones as well. However, he also analysed an "autonomous" orthogenesis occurring irrespectively of the environment. Moreover, Jeannel tried to prove that evolution was mostly regressive. Organisms never form new organs in response to novel functional demands, they just change the ones they had long possessed. Over several hundred million years of their evolution, insects go on modifying their six legs and are unable to invent anything new (Jeannel 1950).

Orthogenesis offered especially satisfactory answers to questions arising in palaeontology, morphology and taxonomy. This does not mean, however, that there were no such questions elsewhere in biology. Orthogenesis also stroke a chord with scientists working in other fields. Some experiments on heredity and variability also provided some support of the idea of directed evolution. Studies of an American zoologist **Charles Otis Whitman** (1842–1910) are especially interesting in this respect, although this has not been fully realised by the advocates of orthogenesis. Whitman conducted studies on heredity and variability of pigeons since 1892 and found abundant evidence in favour of orthogenesis. He tried to create his own "genetics" looking for its principles in the analysis of recapitulation, phylogenetics and orthogenesis. Unfortunately, he died suddenly in 1910 without completing the generalisation of his material. This was done later by his students. Whitman's treatises on pigeons were finally published 9 years after his death (Riddle 1919). By that time genetics had already settled in biology and Whitman's studies, which contradicted it, passed largely unnoticed.

It is common knowledge that Mendel's laws were discovered and rediscovered in studies of the simplest cases of heredity. The number of species opening a window into the world of genetics was rather small. The theory of mutations by de Vries was mainly based on the variations of a single plant species, *Oenothera lamarckiana*. Such an approach was unacceptable for Whitman. He considered the conclusions made by de Vries as extremely hasty and used any occasion to criticise him. Whitman tried to study as much of variable material as it was possible, analysing numerous species and large numbers of characters. His experiments on pigeons included revealing pure lines, crossing, analysis of heredity. It was experiments like this that resulted in the formation of genetics but in Whitman's case they were just a part of a large scale research program. Its frame was continuously extending. Whitman believed that in order to discover the mechanisms of heredity and variability one had to use well-studied groups, i. e. groups with known evolutionary tendencies. Pigeons seemed much better suited for this purpose than *Oenotera*, a plant of unclear, possibly hybrid origin.

Whitman collected a large bulk of data on species and races of pigeons. Comparing them and analysing their individual development, he described the laws of pigeon evolution. Whitman asked question about characters determining differences between species, trying to solve an eternal problem of the evolutionary biology: how to extrapolate the data on variability to the process of the origin of new species? However, formalisation of this procedure turned out to be problematic. Each case required numerous and diverse comparisons. Whitman meticulously carried out the necessary research, constantly increasing the number of characters. Colouration was the most convenient character for the genetic analysis of birds but Whitman did not restrict himself to this. He studied characters related to voice, behavior and fecundity. He compared the differences between races with those between species and those occurring during individual development. He came to the conclusion that all these phenomena were underlain by common laws.

Whitman believed that he proved the existence of orthogenesis, that is, directed evolution determined by the predisposition of organisms to vary in certain directions, not by natural selection. He often criticised Darwin's theory. So, studying the variability of pigeons he looked for "rudiments", initial stages of some selected characters, which could illustrate the direct way of selection but failed to find any. For example, selecting pigeons, it was easy to get a gradual reduction of pigmentation of feathers. This process proceeded from the front backwards, starting from spotted type through the type with two stripes on wings. Pigment never disappeared gradually and evenly over the entire surface, and no other directions were noted.

Despite his criticism of de Vries and Darwin, Whitman claimed that their concepts were compatible with his own. He believed that three concepts—Darwinism, mutationism and orthogenesis—characterised different aspects of evolution. Orthogenetic processes provided material for the play of natural selection, while de Vries' mutations were an example of this process. Similarly to other authors of orthogenetic concepts, Whitman used the analogy between evolution and the growth of crystals (Whitman 1919).

Another "alternative genetics" related with orthogenesis was elaborated by a German scientist **Richard Woltereck** (1877–1944). He also studied the mechanisms of heredity and variability but his conclusions were hardly compatible with the principles of genetics. Since 1905 he conducted experimental research on natural populations of daphnia (Woltereck 1919) and found after several years that the norm of reaction was a hereditary feature (it was Woltereck who coined the term "norm of reaction"—*Reaktionsnorm*). Woltereck considered these data as an indication of the inheritance of phenotypic changes though later he changed his opinion. According to his data, daphnia sometimes showed spontaneous collective hereditary changes in a definite direction. However, his support of the idea of directed evolution was more likely to be associated with his philosophic speculations. Woltereck's interest in philosophy arose when he struggled with depression during the First World War. He was also influenced by his friend, the writer Hermann Hesse. Finally Woltereck almost stopped his biological research and wrote treatises on the entire living world, which were hardly understandable (Harwood 1996).

Several geneticists also expressed their sympathies towards orthogenesis. **Nikolay I. Vavilov** (1887–1943) paid much attention to the studies by Berg and Sobolev. When Berg worked on nomogenesis, Vavilov helped him to obtain the relevant literature. Later he promoted the publication of Berg's book abroad. Berg thought that nomogenesis and the law of homological series were expressions of the same views. Vavilov shared this opinion but was careful never to express it in publications. There are no conclusions contradicting the theory of natural selection in his paper about the law of homological series (Vavilov 1920 cited in Vavilov 1968).

This was partly due to the fact that Vavilov was much more interested in selection and taxonomy of cultural plants than in evolutionary discussions though, evolution and, in particular, orthogenesis attracted him too. He wrote to Sobolev in 1920: "I had recently learnt from Professor L. S. Berg, that in your work on phylogeny of goniatites you express ideas analogous to our conclusions based on studies on plants. I have read your book, and indeed your data on goniatites, the law of parallel variation, is very close to what I have called the law of homological series" (cited in Sobolev 1924, p. 44). Evidently, Vavilov was also familiar with works related to orthogenesis. In the letter to G. S. Zaitsev he mentioned "orthogenetic rows": "We advanced much further than the little that was published in the report. I am afraid we have reached the jungles of the basics of taxonomy of genera and families but then, the essence of the generalisation was directed towards taxonomy. In taxonomy itself, and we spent a lot of time with it, much has become very simple. The essence is not in combinations but in the parallelism of cycles of varying characters. Coniferans do not have much in common with gramineans, nor pumpkins and watermelons with wheat but the cycles of variation are similar in many respects. Such parallelisms are numerous among fungi and among animals, they are easily found in any detailed monograph on any large groups or even genera. Reading *Mutationstheorie*, scrutinizing "mutations", one can see that all of them follow orthogenetic rows." (Vavilov 1977, p. 104).

Later Vavilov tried to downplay the proximity of his studies to "non-Darwinian heresies", very likely because the polemics acquired political undertones. The essence of his ideas did not change however. He continued to study homological series and other regularities of variability. It is interesting that Vavilov approved not only the conclusions of Berg and Sobolev but also those of another critic of Darwinism, **John Willis** (1868–1958). Vavilov shared his "age and area hypothesis" (Willis 1922), which fitted well the concepts of directed evolution and, like nomogenesis, contained a corpus of critical comments against Darwinism. Vavilov's career ended tragically: he was put in prison in 1940 and died there in 1943. His studies, which promised so much, were left unfinished. Several works on homological series were published afterwards by other authors but contained just more of the same facts and added little to the theory itself.

Another Russian geneticist, **Yuri A. Filipchenko** (1882–1930), also sympathised orthogenesis. Some of his conclusions coincided with those of Berg. However, Filipchenko apparently did not want to "choose sides" in evolutionary biology. He believed that evolution occurred under the impact of some internal factors. This seemed to him so evident as not to require any special proof. Yet many biologists were of the opposite opinion, and Filipchenko could hardly ignore this. His ideas about evolution were mainly based on the data on spontaneous origin of mutations. Filipchenko criticised both Darwinism and Lamarckism. From his viewpoint, natural selection was a force that merely sorted out something that had originated for unknown reasons. In his later studies Filipchenko was less sceptical about natural selection but still did not think that it explained everything in evolution. He compared Darwin to Copernicus and thought that biology was still waiting for its Newton. According to Filipchenko, biology could at best explain the mechanisms of microevolution while macroevolution remained out of the reach of our knowledge. (The terms "macroevolution" and "microevolution" were coined by Filipchenko.) Unlike most mutationists, he took into account the fact that although mutations originated spontaneously they were not arbitrary and therefore evolution had to follow definite directions. Numerous parallelisms indicated this, and Filipchenko analysed them carefully (Filipchenko 1924; 1977; 1978).

A German researcher **Ludwig H. Plate (1862–1937)**, who studied zoology, morphology and genetics, supported orthogenesis in his works. This support, however, might seem somewhat inconsistent because Plate tried to unify Lamarckism, Darwinism and orthogenesis, that is, the concepts that were usually in opposition to each other. Initially Plate called his own concept "old Darwinism" to distinguish it from Weismann's Neo-Darwinism. Speculating on natural selection, Plate coined the term "orthoselection" meaning selection in a definite direction. Since selection is by definition an action in a definite direction, the need to supply it with the prefix "ortho- "is questionable. However, this term gained ground later on. Darwinians used it to criticise orthogenesis, claiming that orthogenesis was merely selection in a definite direction rather than a result of spontaneous determinate transformation. Plate himself believed that both orthogenesis and orthoselection existed and that these two processes taken together resulted in orthoevolution. Plate elaborated his own concept concerning genes. He suggested that some genes existed in the nucleus outside chromosomes and that these genes (Erbstock) determined the essential characteristics of organisms, while their transformation in the process of evolution was the mechanism of orthogenesis (Plate 1922, 1925, 1933; Levit and Olsson 2006). This concept did not find any supporters and drifted into obscurity.

Theodosius G. Dobzhansky (1900–1975) was enthusiastic about nomogenesis for a short period of time but then reversed his views (even though he always wrote about Berg with the greatest respect) (Dobzhansky 1980). In general, geneticists tended to support other evolutionary concepts such as mutationism, regressive evolution and Darwinism. Starting from the 1930s, geneticists rarely expressed any support of orthogenesis. They usually did not find in their studies anything that orthogenesis could interpret better. In the 1930s there was probably a single series of genetic studies carried out within the theoretical framework of orthogenesis.

Viktor Jollos (1887–1941), a German geneticist of the Russian origin, found some evidence in favour of orthogenesis in his research of temperature mutations. With the rise of temperature a mutation of eye colour, *dark eosin,* was found in the offspring of wild fruit flies. The subsequent impact of temperature resulted in the following mutations in the same gene: *light eosin*, *yellow*, *ivory* and *white* (Fig. 4.1). No other order of mutations was ever observed. Jollos called them directed mutations implying the tendency to mutate in a definite direction irrespective of adaptation. He believed that this process illustrated series of transformations found in the fossil record such as the trend of evolution of extremities of ungulates. At the same time, Jollos was sceptical of both Lamarckism and Darwinism (Jollos 1931).

The scientific career of Jollos ended tragically. With the establishment of Nazi regime he emigrated, could not find any research position and died soon. His studies of *Dauermodifications* and other "artefacts" were considered suspicious by the orthodox geneticists of that time. The facts of directed mutations found by Jollos were later confirmed but his interpretation was criticised (Harwood 1993). As far as I know, no similar interpretations have been suggested since. It should be mentioned that Michael D. Golubovsky, a geneticist and a historian of science, claimed that he

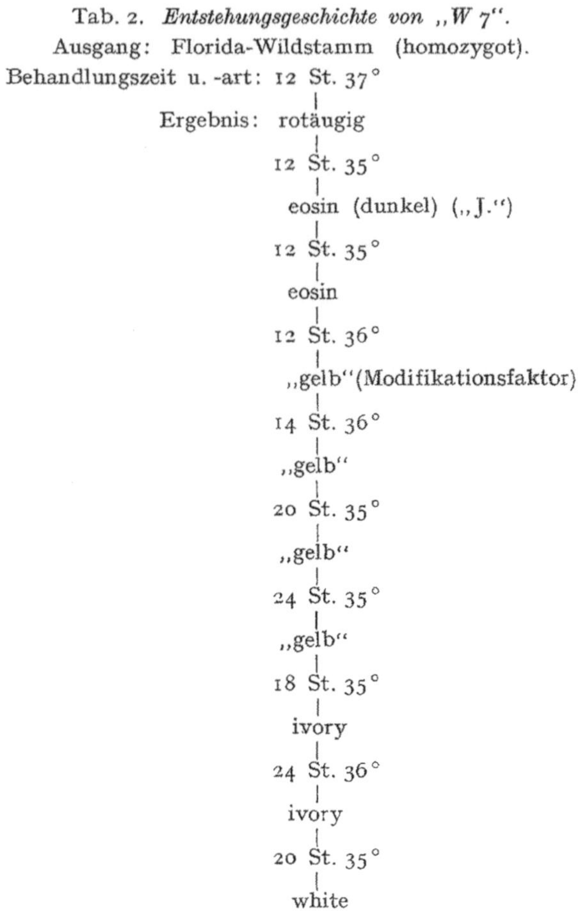
Tab. 2. *Entstehungsgeschichte von „W 7".*
Ausgang: Florida-Wildstamm (homozygot).
Behandlungszeit u. -art: 12 St. 37°
Ergebnis: rotäugig
12 St. 35°
eosin (dunkel) („J.")
12 St. 35°
eosin
12 St. 36°
„gelb"(Modifikationsfaktor)
14 St. 36°
„gelb"
20 St. 35°
„gelb"
24 St. 35°
„gelb"
18 St. 35°
ivory
24 St. 36°
ivory
20 St. 35°
white

Fig. 4.1 Directed mutations discovered by Viktor Jollos

could prove the ten postulates suggested by Berg with the use modern data. However, no instances of "nomogenesis in genetics" are mentioned in his monograph on the history of genetics (Golubovsky 2000). It is not quite clear what kind of proof Golubovsky had in mind but it was apparently not related directly with his own studies on fruit flies.

In the 1920–30s the idea of directed evolution idea surfaced in histology. **Alexey A. Zavarzin** (1888–1945) published a series of papers on parallelisms and evolutionary dynamics of tissues. Their logic has much in common with orthogenesis. Zavarzin noted that anatomic convergences were the initial point of his studies. Supposing that convergences were mostly due to functional reasons, that is, identic functions resulted in identic structures, he started to look for histological parallelisms, or convergences (he used these terms as synonyms), and discovered curious parallelisms in the structure of tissues. Zavarzin studied the nervous tissue in representatives of various phyla and classes. He considered nervous systems as the most suitable for such a research because believed it to be a "frozen" system unaffected by individual functional fluctuations. Indeed, the structure of the nervous system in vertebrates, insects, molluscs and crustaceans turned out to have numerous profound similarities.

The comparison of the spinal cord of vertebrates with the ventral cord of crustaceans seems especially remarkable. It is reminiscent of the famous dispute between Étienne Geoffroy Saint-Hilaire and George Cuvier about the unity of body plan. The dispute was occasioned by a study of cephalopods made by their junior colleagues but extended far beyond the limits of this topic. This discussion was a quintessence of controversies among the theorists of biology of that time. The idea of unity of everything (Geoffroy Saint-Hilaire) was considered progressive because it corresponded well to evolution. On the contrary, the view about stark differences between body plans (Cuvier) was considered as conservative and retrograde (Amlinsky 1955). One of Saint-Hilaire's arguments concerned arthropods: he tried to prove that they were "vertebrates" living inside the vertebra, while Cuvier argued that arthropods and vertebrates had different body plans. Had Zavarzin carried out his studies a hundred years earlier, Geoffroy Saint-Hilaire could have found much support in them. According to Zavarzin, the spinal cord of vertebrates and the ventral cord of insects were different but their microscopic structure was identical. In both cases metameric lateral nerves radiate from the cord by two roots, the tissue is represented by the grey and the white matter etc. Zavarzin (1925) came to the conclusion about the existence of a common plan of the nervous system in the animal world and the general significance of the principle of parallelism in biology (cited in Zavarzin 1986, p. 75.) He noted that this conclusion, similarly to the viewpoint of Geoffroy Saint-Hilaire, seemed paradoxical but the facts it was based on looked convincing.

Zavarzin developed the views that all biological structures originated according to definite laws. In addition to his own data, he recalled the facts cited by Berg about parallelisms of reptiles and mammals, ungulates of Old World and New World and the results of Vavilov's studies on parallelisms in plants. Similarly to Berg, Zavarzin claimed that there was the law of convergence in addition to that of divergence. Polyphyletic forms cannot vary randomly but evolve in definite directions. Therefore convergent characters originate in various parallel evolutionary lineages.

Zavarzin pointed out that his colleagues also observed that histological structures did not fall into phylogenetic limits. To illustrate this, he cited two phylogenetic schemes of arthropods: that of L. Plate based on the structure of eyes and that of B. Handstrem based on the structure of optical centres. These schemes have little in common indicating that different organs can evolve at different rates and relatively independently of each other. A highly developed eye can be accompanied by a primitive nervous support and vice versa. Generally, evolution represents series of types, which are regularly or causally related with each other. Each series can be represented as a ray of a star, while its centre is the primitive condition, in which all structures exist potentially but are indistinguishable morphologically (Fig. 4.2). Zavarzin concluded that "An organism consists of a totality of characters which evolve to a great extent (or even absolutely) independently of each other" (Zavarzin 1928 cited in Zavarzin 1986, p. 92). At the same time, Zavarzin tried to keep away

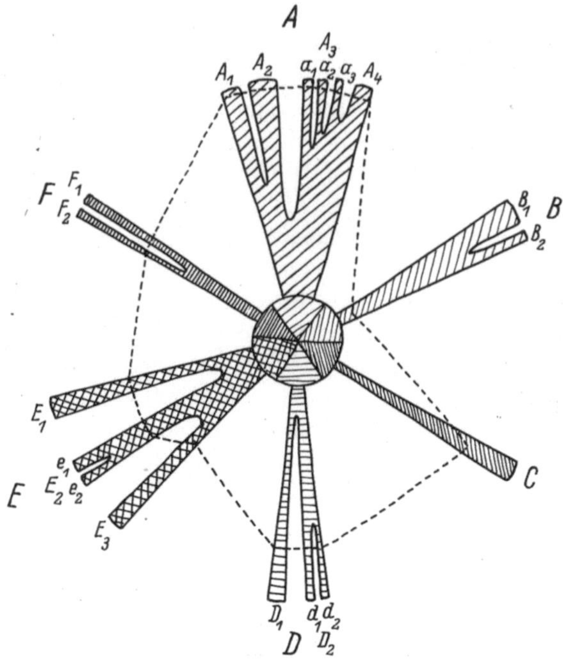

Fig. 4.2 Scheme of the evolution of "types" by Alexei A. Zavarzin (1928)

from the discussions concerning the factors of evolution. Therefore his studies, especially earlier ones, are reminiscent of idealistic morphology. Zavarzin wrote that his viewpoint differed from that of Berg "only in the fact that conclusions about parallelisms and identity of structures are made without any evolutionary approach whatsoever" (Ibid.). His studies agree well with the concepts of directed evolution, as does idealistic morphology.

Zavarzin emphasised the preliminary character of his first generalizations concerning parallelisms, pointing that they were not a fully elaborated theory. He wanted "to come out of some morphological dead ends" using new methods of comparison. Later Zavarzin studied the "evolutionary dynamics of tissues" and supplemented the "principle of parallelisms" with evolutionary analysis. Zavarzin pointed to his own error: he had previously considered evolution of tissues irrespective of the evolution of organisms. However, he still tried to stand aside from evolutionary discussions including the problems related with orthogenesis. To a large extent, this was due to the fact an active support of directed evolution in the USSR of that time was simply unimaginable.

Zavarzin's statements about directed evolution are rather contradictory and can be interpreted in different ways. It seems expedient to cite some of his writings here.

"There is a certain directionality in phyletic evolution, which is the only way to explain the parallel rows.

We would look in vain for any such directionality in the evolution of entire organisms. Phylogenetic development of an organism is determined by causes of a different nature, by more complicated causes whose general interactions have been compiled into the most logical Darwin's theory in its dialectical interpretation.

This theory copes splendidly with all the facts in which the parallelisms and directionality are absent and always encounters difficulties when it is necessary to explain convergences, which, as we have seen, abound in the evolution of tissues where similarities evidently dominate.

Should we draw from that a conclusion that Darwin's theory is incorrect or inapplicable to histological facts, as the few authors interested in the application of the evolutionary doctrine to the histological material have been doing up to now?

Surely not. However, we must draw a conclusion that the fact of directionality of phylogenesis of tissues should find its own evolutionary explanation." (Zavarzin 1934 cited in Zavarzin 1986, p. 115).

This indicates that Zavarzin agreed with the opinion presented by the advocates of orthogenesis but did not want to present it as an alternative to the natural selection theory. Afterwards he found an explanation of the mechanism of directionality of the "phylogenesis of tissues". He believed that while the internal environment of organisms "remains practically unchanged, the living substance cannot help changing because otherwise it cannot remain alive... […] As function and the general biological constant of the internal environment of all animals remain relatively constant, so its differential realisation mostly remains in the same forms. The fact of the existence of parallel rows and the directionality of evolution of tissue system stems from this." (Ibid., p. 118).

Further analysis of the "evolutionary dynamic of tissues" resulted in conclusions identical with those of orthogeneticists. Zavarzin reworded their main idea in a more cautious manner as far as unknown internal mechanisms of evolution were concerned: "If one attaches exclusive importance to the role of function in differentiation, the differentiating system would only get the role of an inert pliable mass moulded in various ways by the external conditions. In such a conditionalistic conception, which does not take into account the specific features of the differentiating system itself, i. e. the specific features of the living matter, the explanation misses the mark, the very thing that needed explanation—the essence of the differentiation process—slips and only conditions remain. Thus, the division of labour alone is evidently insufficient to explain differentiation. At the same time, such an explanation, due to its seeming simplicity and persuasion, often impresses so much that one calms down and does not go deeper into the matter.

A reverse explanation from form to function corresponds better to the facts but, if taken unconditionally, necessitates the recognition of an immanent purposefulness in evolutionary processes, which is unacceptable from the materialistic point of view serving as the basis of the present work and being, of course, the only right one in the research of natural phenomena.

The insufficiency of both explanations given above lies in their one-sidedness, since each of them takes into account only one aspect and does not explain the phenomenon in its entirety." (Ibid., p. 137).

Zavarzin thought that he succeeded in demonstrating that the "directionality and parallel rows do not contain anything non-material or teleological but are the natural consequence of development (self-motion) of the living matter under conditions of its earthly existence provided the quality of organic unity is removed" (Ibid., p. 130). So, despite his evident reluctance to criticise the theory of natural selection, Zavarzin repeatedly came to "orthogenetic" conclusions. Elegant and expressive wordings of compromises are scattered throughout his writings but a perfect command of the language can hardly be considered as a solution to the problem. The dilemma "orthogenesis vs. Darwinism" still existed, and the analysis of parallelisms remained the field where the advocates of orthogenesis found their strongest arguments.

Several biologists chose orthogenesis as a theoretical frame for the presentation of their results. Describing material on variation and evolution of organisms, they did not refer to their selective of adaptive background but considered them as manifestation of yet partially obscure evolutionary laws. The existence of some predisposition in evolution was usually considered to be self-evident. The authors of such studies partly supported orthogenesis but tried to avoid collisions with Darwinism by choosing to speculate in other terms and on other topics than the authors of the best known evolutionary works.

This kind of orthogenesis flourished in research dealing with the organisms' ability to synthesise organic substances. This phenomenon may be referred to as biochemical or physiological evolution. The existence of such evolution was first suggested in the late nineteenth century by **Frederick Gowland Hopkins** (1861–1947), an English biochemist who studied vitamins. In the 1910–1920s several "tendencies" of biochemical/physiological evolution of bacteria, protists and fungi were discovered (Lwoff 1943). In the 1920s **Andrey V. Blagoveshchensky** (1889–1982) carried out an extensive research on the concentration of alkaloids in plants and showed that their concentration increased gradually in many phyletic lineages of plants. The concentration of alkaloids and other cyclic substances seemed to indicate the senescence of taxa. The greatest amount of cyclic substances were found in the plants with primitive features, which seemed to have frozen in their morphological evolution and were the relics of groups which had flourished in the past. Blagoveshchensky tried to explain these facts on the basis of principles of energy transformation. He believed that in the beginning of the evolution of a plant taxon, its representatives synthesised aliphatic substances (rich in energy) and then increasingly more cyclic substances (which are less energy-rich). In this way, evolution strived towards the most energy-favourable condition. At the same time, cyclic substances are a dead end for chemical reactions in an organism because they can only be accumulated and not used. An organism with much of these substances was also an evolutionary dead end. Blagoveshchensky thought that evolution was a directed, or orthogenetic, process from the viewpoint of energy, although he did not support orthogenesis outright and did not discuss the difficulties of the natural selection theory (Blagoveshchensky 1925). Later he repeatedly confirmed his viewpoint (see Blagoveshchensky and Alexandrova 1974).

Physiological evolution was analysed in most detail by a French scientist André Lwoff. He presented the main points of his concept in 1930 and published a monograph on this subject in 1943 (Lwoff 1943). Lwoff focused on the "capacity for synthesis" in many groups of organisms. He showed that the capacity to synthesise the necessary substances decreased with the increasing complexity. The organism became progressively more dependent on external sources of complex organic substances, especially amino acids and vitamins, and the so-called physiological degradation occurred. Lwoff revealed several tendencies in the evolution of biochemical characters such as the loss of pigments in protists, the loss of the capacity to synthesise alanine and aneurine etc.

In relation with these facts, Lwoff analysed orthogenesis, adaptation, progress and other phenomena usually discussed in general evolutionary treatises. However, he did not want to step outside the framework of the phenomena under study and pointed out questions rather than suggested answers. He wrote about adaptation in the following vein: in discussions of adaptations one camp is represented by vitalists, like Cuénot, which ascribe a finalistic sense—that is, the correspondence to biological purposefulness which had originated due to unknown reasons—to every particularity of an organism. The other camp is represented by mechanists, like Etienne Rabaud, who deny adaptation. According to Rabaud, if an organism is alive, this means that it is adapted, and all speculations about adaptation are fantasies. Lwoff did not want to join either camp though he noted that there were several groups of facts, which, though usually not discussed, testify against vitalists: a) the existence of meaningless or harmful characters and b) several species, genera, families and orders became extinct or border on extinction.

Analysing his material, Lwoff pointed out the complexity and inconsistency of the data on progress and adaptation. On the one hand, there are negative, i.e. non-adaptive, changes in physiological evolution. For instance, having lost chlorophyll, an organism has become more dependent on the source of organic nutrition. This means that it has acquired a non-adaptive character. On the other hand, there are arguments in favour of an opposite interpretation of physiological evolution. For instance, the flagellate *Polytoma ocellatum* has lost the capacity to synthesise thiasol, which increases its dependence on specific food sources. From a certain viewpoint, this event is regressive. However, it can also be considered as progressive: any synthesis requires energy and since this organism can do without energy expenditure for the synthesis of complex substance, this strategy is profitable in respect of thermodynamics. This means that, paradoxically, regressive evolution can be considered as adaptation. Still, Lwoff preferred the term "regressive orthogenesis" to adaptation, specialisation etc. He considered physiological/biochemical evolution as resulting from spontaneous transformation rather than selection of small variations. "Regressive orthogenesis" did not mean that Lwoff did not search for evidence of progressive physiological evolution. He assured the readers that he did try to find it and failed. Evolution of the capacities to synthesis was always regressive.

Lwoff wrote that after these ideas occurred to him for the first time, he discussed them with his colleagues (*maîtres*) working in the field of botany and zoology but found a lack of understanding. The *maîtres* did not exactly criticise him—there was nothing in Lwoff's data they could possibly object to. Rather, they were annoyed and tried to ignore these data. In contrast, Lwoff did not think that physiological evolution was something essentially different from "normal" one in respect of mechanisms. From his viewpoint, both evolutions have a common source and represent different aspects of change of hereditary material, which varies not randomly but in definite directions.

Almost simultaneously with Lwoff, "orthogenetic evolution of biochemical systems" was described by a Belgian scientist **Marcel Florkin** (1900–1979). He described several evolutionary tendencies in animals and humans. Colloid osmotic pressure of proteins increases with the rise of the level of organisation irrespective of ecological conditions, the content of carbohydrates in proteins decreases, the oxygen capacity of blood increases, the proportion of extracellular digestion increases etc. In some cases, e.g. in the evolution of decomposition of purines, orthogenesis was regressive rather than progressive. The reaction chain during decomposition of purines follows the same pattern in all living beings, but as the level of organisation rises it shortens irrespective of adaptation. The purines are decomposed down to the end (ammonium) in lower organisms but down to more complex substances in higher organisms. Florkin considered this phenomenon as regressive biochemical orthogenesis (Florkin 1947). Tendencies of this kind were revealed in different phylogenetic lineages. Florkin concluded that biochemical systems seemed to evolve in the directions determined by the common biochemical basis of all animals (Florkin 1947).

To sum up, many authors declared their support of orthogenesis in the 1900s–1930s. Some of their arguments seem convincing. Gradually, however, the ideas of modern Darwinism became more popular. In general, studies that agreed with the natural selection theory such as genetics of populations, experimental studies of geographic races and "new systematics" developed in the 1930s more actively.

References

Amlinsky IE (1955) Geoffroy Saint-Hilaire and his struggle against Cuvier (Zhoffrua Sent-Iler i ego bor'ba protiv Kyuv'e). AN SSSR, Moscow

Banta AM (1907) The fauna of Mayfield cave. Carnegie Institution, Washington

Beurlen K (1930) Comparative evolution (Vergleichende Stammesgeschichte). Borntraeger, Berlin

Blagoveshchensky AV (1925) On the question of directionality of the process of evolution (K voprosu o napravlennosti protsessa evolyutsii). Byulleten' Sredneaziatskogo Gosudarstvennogo Universiteta 10:17–33

Blagoveshchensky AV, Alexandrova EG (1974) Biochemical basis of the phylogeny of higher plants (Biokhimicheskie osnovy filogenii vysshikh rasteniy). Nauka, Moscow

Blyakher LYa (1962) An essay of the history of animal morphology (Ocherk istorii morfologii zhivotnykh) AN SSSR, Moscow

Blyakher LYa (1976) Problems of animal morphology. Historical essays (Problemy morfologii zhivotnykh. Istoricheskie ocherki). Nauka, Moscow

Bocking S (1988) Alpheus Spring Packard and cave fauna in the evolution debate. J Hist Biol 21(3):425–456

Cei G (1946) Parallel orthgenesis and degradation of organs of sight in Splacidi (Orthogenesi parallela e degradazione degli organi della vista negli Spalacidi). Monitore Zoologico Italiano, Anno LV, 7–12:69–88

Dacqué E (1915) Palaeontology, systematics and evolutionary teaching (Paleontologiya, sistematika i evolyutsionnoe uchenie). In: Vagner VA (ed) New ideas in biology. Collection of papers 8 (Novye idei v biologii. Sbornik 8). Obrazovanie, Petrograd, pp 40–73
Dacqué E (1921) cited in Levit GS, Meister K (2006) The history of essentialism vs. Ernst Mayr's "Essentialism Story": a case study of German idealistic morphology. Theory Biosci 124(2006):281–307
Davitashvili LSh (1948) The history of evolutionar palaeontology from Darwin to our days (Istoriya evolyutsionnoy paleontologii ot Darvina do nashikh dney). AN SSSR, Moscow, Leningrad
Dean B (1904) Evolution in a determinate line as illustrated by the egg-cases of Chimeroid fishes. Biol Bull 7(2):105–113
Dobzhansky Th (1980) The birth of the genetic theory of evolution in the Soviet Union in 1920s. In: Mayr E, Provine W (eds) The evolutionary synthesis: perspectives on the unification of biology. Harvard Univ press, Cambridge (Mass), London, pp 229–242
Dollo L (1893) The laws of evolution (Les lois de l'évolution). Bulletin de la société belge de géologie, de paléontologie et hydrologie 7:164–166
Fage L (1931) Underground evolution and its determinism (L'évolution souterraine et son déterminisme). Archives de zoologie expérimentale et générale 71:99–291
Filipchenko YuA (1924) On parallelism in the living nature (O parallelizme v zhivoy prirode). Uspekhi sovremennoy biologii 4:242–257
Filipchenko YuA (1977) Evolutionary idea in biology: a historical review of evolutionary teachings of the 19th century (Evolyutsionnaya ideya v biologii: *istoricheskiy Obzor evolyutsionnykh ucheniy XIX veka).* Moscow
Filipchenko YuA (1978) Variability and he methods of its study (Izmenchivost' i metody ee izucheniya), 5th edn. Nauka, Moscow
Florkin M (1947) Biochemical evolution (L'évolution biochimique). Desoer, Liège
Gaydukov NM (1925) On the phylogenetic system of angiosperms in connection with serum diagnosis, complications and convergencies (O filogeneticheskoy sisteme skrytosemennykh v svyazi s serodiagnostikoy, komplikatsiyami i konvergentsiyami). Zapiski Belorusskogo instituta sel'skogo i lesnogo khozyaystva, 8:67–81
Gaydukov NM (1926) On convergence, complications and phylogenetic system of schizophytes and algae (O konvergentsii, komplikatsiyakh i filogeneticheskoy sisteme drobyanok i vodorosley). Russkiy arkhiv protistologii 5(3–4):269–298
Golubovsky MD (2000) The age of genetics: evolution of ideas and notions (Vek genetiki: evolyutsiya idey i ponyatiy). SPbFIIET RAN, St Petersburg
Haffer J, Rutschke E, Wunderlich K (2000) Erwin Stresemenn (1889–1972). Life and work of a pioneer of scientific ornithology (Erwin Stresemenn [1889–1972]. Leben und Werk eines Pioners der wissentschaftlichen Ornithologie). Acta Historica Leopoldina 34, Deutche Akademie der Naturforscher Leopoldina, Halle (Saale)
Harwood J (1993) Styles of scientific thought: the German genetics community, 1900–1933. Univ of Chicago press, Chicago
Harwood J (1996) Weimar culture and biological theory: a study of Richard Woltereck (1877-1944). Hist Sci 34(3):347–377
Jeannel R (1950) The march of evolution (La marche de l'évolution). Edition de muséum, Paris
Jollos V (1931) Experimental origin of mutants and their significance for the problem of evolution (Die experimentelle Auslösung von Mutanten und ihre Bedeutung fur das Evolutionsproblem). Naturwissenschaften 19:171–177
Kowalevsky WO (1875) The osteology of two extinct species of ungulates (Osteologiya dvukh iskopaemykh vidov iz gruppy kopytnykh). Izvestiya obshchestva lyubiteley estestvoznaniya, antropologii i etnografii, 16:1–59
Lessertisseur J (1959) Darwinism and palaeontology (Darwinisme et paléontologie). In: Revue générale des sciences pures et appliquées, vol 66, pp 153–160
Levit GS, Olsson L (2006) Evolution on rails: mechanisms and levels of orthogenesis. Ann Hist Phil Biol 11:97–136

Lwoff A (1943) Physiological evolution. A study of losses of function in microorganisms (L'évolution phisiologique. Étude des pertes de fonction chez les microorganisms). Actual Sc Industr 970
Nazarov VI (1991) The teaching about macroevolution. On the ways towards a new synthesis (Uchenie o makroevolyutsii. Na putyakh k novomu sintezu). Nauka, Moscow
Neumayr M (1919) The roots of the animal kingdom. Introduction into the science of the origin of animals (Korni zhivotnogo tsarstva. Vvedenie v nauku o proiskhozhdenii zhivotnykh). Izdatel'stvo M. and S. Sabashnikovykh, Moscow
Plate L (1922) General zoology and evolution (Allgemeine Zoologie und Abstammungslehre). Gustav Fischer Verlag, Jena
Plate L (1925) Evolution: facts, theories, objections and conclusions in brief presentation (Die Abstammungslehre: Tatsachen, Theorien, Einwände und Folgerungen in kurzer Darstellung). Gustav Fischer Verlag, Jena
Plate L (1933) Science on heredity: with special attention to evolution and humans. V. 2. Sexuality and general problems (Vererbungslehre: Mit besonderer Berücksichtigung der Abstammungslehre und des Menschen. Bd. II Sexualität und Allgemeine Probleme). Gustav Fischer Verlag, Jena
Riddle O (ed) (1919) Posthumous works of Charles Otis Whitman. Carnegie Institute, Washington
Rieppel O (2012) Karl Beurlen (1901–1985), nature mysticism, and Aryan paleontology. J Hist Biol 45:253–299
Sarasin P, Sarasin F (1899) Land molluscs of Celebes (Die Landmollusken von Celebes). C. W. Kreidel, Wiesbaden
Severtsov AN (1914) Present-day tasks of evolutionary theory (Sovremennye zadachi evolyutsionnoy teorii). Nauka, Moscow
Simpson GG (1944) Tempo and mode in evolution. Columbia Univ. Press, New York Russian edition
Simpson GG (1948) Tempy i formy evolyutsii. Gosudarstvennoe izdatel'stvo inostrannoy literatury, Moscow
Simpson GG (1950) Orthogenesis and the synthetic theory of evolution (L'orthogenèse et la théorie synthétique de l'évolution). Paléontologie et transformisme. Albin Michel, Paris, pp 123–168
Sobolev DN (1924) The fundamentals of historical biogenetics (Nachala istoricheskoy biogenetiki). Gosudarstvennoe izdatel'stvo Ukrainy, Simferopol
Steinmann G (1908) Geological Foundations of the Evolution theory (Die geologischen Grundlagen der Abstammungslehre). Engelmann, Leipzig
Stresemann E (1975) Ornithology from Aristotle to the present. Harvard University Press, Cambridge, MA/London
Taylor PD (2002) W. D. Lang, Orthogenesis and the evolution of Cretaceous cribrimorph bryozoans. In: Patrick N, Wyse J, Jones MES. Annals of Bryozoology. International Bryozoology Association, Dublin, pp 275–299
Vavilov NI (1920) cited in Vavilov NI (1968) The law of homological sereis in hereditary variability (Zakon gomologicheskikh ryadov v nasledstvennoy izmenchivosti). In: The classics of Soviet genetics (Klassiki sovetskoy genetiki). Nauka, Leningrad, pp 9–51
Vavilov NI (1977) Letters of N I Vavilov to G S Zaitsev Priroda 4:102–115
Whitman ChO (1919) Orthogenetic evolution in pigeons. In: Riddle O (ed) Posthumous works of Charles Otis Whitman. Carnegie Institute, Washington
Willis JC (1922) Age and area. Cambridge Univ press, Cambridge
Woltereck R (1919) Variation and speciation (Variation und Artbildung). Verlag von A. Francke, Bern
Zavarzin AA (1925) cited in Zavarzin AA (1986) Works on the theory of parallelisms and evolutionary dynamics of tissues (Trudy po teorii parallelizma i evolyutsionnoy dinamiki tkaney). Nauka, Leningrad

Chapter 5
Orthogenesis and the Modern Evolutionary Synthesis

Abstract The modern evolutionary synthesis, also referred to as the modern Darwinism, was formed in the 1930–40s and has dominated biology ever since. Its success is partly due to the energetic claims of its authors that they had solved all the main problems of the evolutionary biology. The modern Darwinism was formed over a long period of time by many scientists. This circumstance was both its weakness and its strength. On the one hand, the vagueness of its main postulates left it open to criticism. On the other hand, so much was written about this theory that most critical remarks could be addressed by finding a reference to some concession, exception or particular case.

Keywords Evolutionary synthesis · Modern Darwinism · Criticism · Period of formation · Vagueness

The modern evolutionary synthesis, also referred to as the modern Darwinism, was formed in the 1930–40s and has dominated biology ever since. In general, it has successfully resisted numerous attempts at revision, which were prompted by new discoveries. The modern Darwinism was criticised quite actively in the 1970s, especially by saltationists, who claimed to have achieved a new synthesis (Gould 1977, 1982; van Waesberge 1982), and also by the advocates of Lamarckism, neutralism and orthogenesis. At the same time, the best known alternatives to the modern synthesis could not offer any essentially novel evolutionary ideas. This is especially true of the seemingly new saltationistic concepts, which were not actually new at all. Ideas about "adaptive mutations" or neutral biochemical changes in evolution were indeed based on new facts but they did not disagree with the traditional views on evolution, and their authors often stressed this (Kimura 1983; Stahl 1988). They showed that the process of evolution was more complex than considered by the early versions of the modern synthesis in the 1940s and specified some details but did not suggest any alternative to the modern synthesis.

Any author writing about evolution can put a claim to the creation of a new synthesis. In addition to the familiar examples cited above, numerous "new syntheses" can be found in less known sources. A conference *Towards a new synthesis in evolutionary biology* took place in the Czech republic in 1987, a Doctor thesis about a

I. Popov, *Orthogenesis versus Darwinism*,
https://doi.org/10.1007/978-3-319-95144-7_5

new "ecological-genetic" synthesis was defended by Ya. Gall in the USSR (Gall 1988), the book by V. Nazarov above cited (1991) was subtitled "On the ways towards a new synthesis" etc. It is difficult to prove that a new synthesis is really new yet attempts to achieve it never not cease. In the last decade it was argued again that a new, extended or expanded synthesis was necessary because of the emergence of new disciplines and the accumulation of new data (Pigliucci 2007; Müller 2007). Yet again it was argued back that the new disciplines and the new data could be very well included in the previous synthesis (Laland et al. 2014). At the same time, as long as any new, extended or expanded synthesis remains Darwinian, it does not question the leading role of natural selection in evolution. Therefore, it is largely irrelevant in the context of the "orthogenesis versus Darwinism" discussion.

Therefore, the last decades can be considered as a relatively continuous period in the history of biology dominated by the modern evolutionary synthesis. It is difficult to say when exactly it started. The formation of the modern synthesis was gradual, and took time. The publication of Dobzhansky's *Genetics and the origin of species* (1937) is usually considered as the starting point. This book generalised the data on genetics, especially the genetics of populations, in terms of the natural selection theory. Somewhat later the principles of natural selection were applied to the data from genetics and other disciplines in several treatises published in the USA and Great Britain: a collection of papers entitled *New systematics* edited by Julian Huxley (1940), Huxley's own book *Evolution. The modern Synthesis* (1942), *Systematics and the origin of species from the viewpoint of zoologist* by Ernst Mayr (1942) and *Tempo and mode of evolution* by George Gaylord Simpson (1944). The title of Huxley's book was the banner under which Darwinists gathered. Yet for a long time the "synthesis" remained not quite "synthetic". The first works were based on zoological material, and the synthesis was accused of "zoological chauvinism" (Stebbins 1980). The misbalance was partly corrected by **George Ledyard Stebbins** (1906–2000) in 1950 (Stebbins 1950). By that time the movement was joined by German and Russian scientists. For obvious reasons, they could not collaborate much with the foreign colleagues in the 1940s but modern Darwinism had been developing in the USSR by **Yefim I. Lukin** (1904–1999) and **Ivan I. Schmalhausen** (1884–1963) (Schmalhausen 1938, 1939, 1946; Lukin 1940) and in Germany by **Nikolai W. Timofeeff-Ressovsky** (1900–1981), **Bernhard Rensch** (1900–1990) and **Gerhard Heberer** (1901–1973) (Timofeeff-Ressovsky 1939; Rensch 1943; Heberer 1943).

Analysing the formation of the evolutionary synthesis in 1980s, Ernst Mayr named only six "architects": Dobzhansky, Huxley, Simpson, Rensch, Stebbins and himself. Other scientists, in his opinion, just prepared the ground for it (Mayr 1982). Such an interpretation may be questioned, especially outside the English-speaking world, but seems justified if we consider the synthesis not simply as a scientific concept but as a manifestation of an organised community of evolutionists. The efforts of the architects were focused on eradicating disagreements in their ranks. They tried to downplay contradictions although they had different views on several significant points of the evolutionary concept. So, Mayr and Simpson considered each other's viewpoints on the biological species as "non-biological" (Hull 1988).

However, this did not prevent them from organising (together with Dobzhansky, who had his own, and different viewpoint on the species problem) the "New York circle" (Cain 1993), which occupied the central position in the modern synthesis, in the Society for the Study of Evolution and in the *Evolution* journal. There were also disagreements between other architects as well the scientists who "prepared the ground" for the modern synthesis. An American historian William Ball Provine, who studied the formation of the modern synthesis, came to the conclusion that each author considered his contribution to be belittled and consequently felt the need to emphasise his own merits (Provine 1987). Mayr stressed the role of naturalists and, by implication, his own; Sewall Wright (1889–1988) disputed with Mayr defending the theoretical genetic of populations; Huxley claimed that he was the sole true author of the modern synthesis etc. Nevertheless, all these authors considered each other's works as parts of the same trend in evolutionary biology.

This means that, according to Mayr, the formation of the modern evolutionary synthesis took at least 13 years. Formally, it dates back to 1950 when Stebbins' book was published. However, the modern synthesis remained incomplete for years after that. The misbalance of material was not overcome: genetics prevailed. In the first issue of *Evolution*, the central mouthpiece of the modern synthesis, 96 pages out of 112 were occupied by genetic papers, including 62 papers based on studies of fruit flies, while papers on ornithology occupied 6 pages and those on palaeontology, 10 pages. A similar proportion was retained in the following issues. One of the readers mocked Ernst Mayr, who was the journal's editor, advising him to rename it *Genetics and Evolution*. Such critical remarks were numerous, and Mayr promised to balance the material but did not succeed. He asked scientists from other biological fields to submit papers on evolution but they either would not do it or wrote papers inacceptable for the "architects" (Cain 1994).

The "zoological chauvinism" was not eliminated, either. There was much less botanical than zoological material into the modern synthesis. It is interesting that facts from botany were often used to illustrate exceptions. Even Stebbins, which wholeheartedly supported Dobzhansky and Mayr, often cited botanical examples contradicting their concepts such as the evolutionary role of hybridization, the possibility of saltations, instances of non-geographic speciation, exceptions from the concept of biological species etc.

An impression of incompleteness produced by the modern synthesis is also due to disagreements among the main theorists, who, in addition, sometimes changed their opinion on some evolutionary factors. A collective monograph Evolutionary synthesis was published only in 1980 (Mayr and Provine 1980). In the same year Nikolai N. Vorontsov (1934–2000) published a synopsis of modern Darwinism (Vorontsov 1980). Yet both these works contained contradictory concepts. Their only common feature was the acceptance of the natural selection principle as the most significant factor of evolution. This circumstance was both the weakness and the strength of the modern synthesis. On the one hand, the vagueness of its main postulates left it open to criticism. On the other hand, the architects of the theory wrote so much on the topic that most critical remarks can be addressed by finding a reference to some concession, exception or particular case.

The success of the modern synthesis is partly due to the energetic claims of its authors that they had solved all the main problems of the evolutionary biology and that their views were generally accepted. This position is convenient; whether it is true is another matter. To begin with, hardly any French scientists participated in the formation of the evolutionary synthesis. The architects stressed this fact, implying that by declining to participate in the formation of modern Darwinism French scientists distanced themselves from evolutionary biology in general (*The evolutionary synthesis*, 1980, p. 309). However, the long-standing tradition of evolutionary studies in France was not broken in the 1930–50s. Even the research directions leading to the modern synthesis developed, though they were not very popular. Ingenious experiments by the French geneticist **George Teissier** provided some of the most compelling evidence in favour of modern Darwinism. He managed to model a population of *Drosophyla* under laboratory conditions and to study selection processes in detail. In particular, he showed that mutant genes, even if they were lethal, did not disappear but remained fixed in a population (Boesiger 1980; Nazarov 1981).

Several French scientists indeed denied evolution at that time. In the introduction and the conclusion to an encyclopaedia published in 1937, Paul Lemuan stated resolutely that the fact of evolution was yet unproven. The authors of the encyclopaedia entries, however, were of a different opinion, and protested in 1938 (Boesiger 1980). Works on evolution were published in France at the same time as the treatises of the architects of the modern synthesis. In France, however, evolution was not dressed in the garb of Darwinism. When the centenary of Darwin's book was celebrated in 1959, French biologists were unenthusiastic. Apparently they failed to see anything significantly new in the writings of Darwin as compared to the works of Lamarck, Buffon or even Anaximander or Empedocles. Teissier pointed this out, noting that French biologists were Lamarckians rather than Darwinians (Teissier 1962). However, Lamarckism did not mean exercises, habits or other anachronisms in this case. It was used as a symbol of the evolutionary idea.

A deeper analysis seems to indicate that French evolutionists simply did not consider themselves as belonging to any trend in the evolutionary biology. In most cases they analysed various theories revealing weaknesses and stressing unsolved problems. There were a dozen authors or so who published general evolutionary treatises in parallel to the architects of the modern synthesis. Some of them expressed ideas that seem archaic even by the standards of that time such as chemical Lamarckism of Paul Wintrebert (Wintrebert 1952, 1962) or somatic reduction by Rene Jeannel (Jeannel 1950). Others, however, were certainly not anachronistic (e.g. Rostand 1962). It is difficult to name most influential spokesman for evolutionary biology in France, that is, a scientist who stood in a position similar to that of the architects of the evolutionary synthesis in English-speaking countries. **Albert Vandel** (1894–1980) and **Pierre-Paul Grassé** (1895–1985) are probably the best candidates for this role. They published several books on evolution expressing their ideas with firm persuasion (Vandel 1954, 1958, 1964, 1968; Grassé 1962) and are the best-cited authors in other French writings on evolution. Both Vandel and Grassé supported the idea of directed evolution and criticised Darwinism. Besides, the authors belonging to the previous generation of evolutionists, Lucien Cuénot and Emile Guyénot, continued to publish until the 1960s, and their works also stood in sharp contrast to the modern synthesis (Cuénot 1951; Guyénot 1961). This trend might have prompted an alternative way of the evolutionary biology development but was gradually extinguished in the 1970s.

In the 1970s things changed. While criticism of Darwinism continued, modern evolutionary synthesis also stroke a chord among French scientists. This situation is reminiscent of the nineteenth century: *The origin of species* was neglected in France for a long time, finding its way into French biology gradually and with numerous reservations. In the 1960s a group of evolutionists was formed in Dijon under the leadership of Henri Tintant, which was loyal to the modern synthesis. They continue to publish on evolution. They do not fully support the modern synthesis, however. Rather, their works are a compromise between the modern synthesis and the old scepticism. The Dijon authors are much less self-confident than the architects of the evolutionary synthesis. They tend to put questions rather than make pronouncements and try to find agreement over unclear points. For instance, they reject the "mystical" orthogenesis but admit the existence of constraints on variation and their significance in evolution; they pay a tribute to the modern synthesis but point out the need for a new one etc. (Devillers and Chaline 1989).

It has been suggested that not only in France but generally in the non-English-speaking world the modern synthesis is not so widespread as it would seem (Bowler 2003). Several German morphologists and paleontologists, notwithstanding the advances of the authors working within the framework of modern Darwinism (B. Rensch, N. W. Timofeeff-Ressovsky, G. Heberer), remained sceptical about the natural selection theory although they did not object to it actively (Reif 1983). A similar situation was observed in Sweden. Idealistic morphology has followed its own course there until recently. As the modern synthesis was being developed, prominent Swedish scientific presented alternative concepts that would have curdled the blood of any modern Darwinian (Olsson 2003).

Thus, the claim that the modern synthesis was generally accepted were hardly justified. However, its authors tended to represent all arising difficulties as minor troubles unworthy of serious consideration. A similar mood had existed in the ranks of active Darwinists before, and used to irritate their opponents extremely. The comments of **Otto Kleinschmidt** (1870–1954) on Darwinism are demonstrative in this regard. Kleinschmidt, who was in his time a well-known anti-evolutionist, published critical comments on evolution during the First World War comparing the writings of Darwinists with military reports of Britain and France: both hushed up failures and exaggerated minute successes (Hossfeld 2003). As it turned out, he was wrong about the military successes but at least partly right about evolutionary biology. However impressive the achievements of the modern evolutionary synthesis might have seemed, alternative concepts, including orthogenesis, continued to emerge and develop, to the great bewilderment of the Darwinists. Similarly to the previous period, in the second half of the twentieth century orthogenesis either emerged anew or was recalled and reworked in modern terms.

Modern Darwinists could not ignore orthogenesis, and mention of several kinds of orthogenetic concepts can be found in their publications. Yet these discussions were often reduced to a minimum. It was announced that orthogenesis was about orthoselection (selection in a definite direction), while all non-Darwinian concepts were considered idealistic. The book *Directionality of evolution* by Alexey S. Severtsov is written in this vein. Though the title seems to suggest the opposite, the author claimed that the natural selection theory was generally accepted and, further, that "it is also widely accepted that selection is not only the moving but also the directing factor of evolution" (Severtsov 1990, p. 8). Evidently, further discussion of orthogenesis was meaningless from such a point of view.

In this context, the notion of directed evolution, if mentioned at all, was used in a different sense than that of the advocates of orthogenesis. It was used descriptively, and its meaning was distorted. Some directions can be revealed in any evolutionary or developmental trend. Ignoring the details of the mechanisms of transformation and just calling evolution "directed", one can evade the discussion of the arguments presented by orthogenesis. The analysis of "directionality" becomes merely a description of the history of the origin of a group of organisms or a biological structure. "Directionality" of the evolution of land vertebrates was described in such terms (Khozatsky 1973, 1981). A change of terminology was the only concession to non-Darwinians in this case.

Several theorists of the modern Darwinism simply never mentioned orthogenesis though they were evidently familiar with it (e.g. Mayr 1942, 1963, 1970). Sometimes they employed the method of exceptions: what is the use of looking for another explanation of evolution if a good one is available? Such an approach is prominent in the works of Bernhard Rensch. He was considered somewhat of an expert in philosophy and directed evolution among the fellow architects of the modern synthesis. Composing his contribution to it, he generalised an abundant material on morphology and taxonomy. He analysed—from a zoologist's viewpoint, as he claimed—the irreversibility of evolution, an increasing specialisation, orthogenetic series in the evolution of organs etc. In every case he pronounced that the theory of natural selection explained these phenomena perfectly well and there was no need for any other concepts. "With this negative statement the absence of autonomous forces of development is of course not proven. We have however no reason to accept the existence of some completely unknown factors, unless we are forced to do so." (Rensch 1954, p. 342). Criticising orthogenesis, Rensch noted that it had never been popular among zoologists. Apparently he considered this as a sort of an argument, though scientific problems can hardly be solved by the majority vote.

The use of the principle of exclusion and "majority vote" by the authors of the modern evolutionary synthesis are in stark contrast with the works of Darwin himself, who was resolutely opposed to such tricks. In his autobiography he recalled the lesson he had learned from the use of this methodology. "Having been deeply impressed with what I had seen of the elevation of the land in S. America, I attributed the parallel lines to the action of the sea; but I had to give up this view when Agassiz propounded his glacier-lake theory. Because no other explanation was possible under our then state of knowledge, I argued in favour of sea-action; and my error has been a good lesson to me never to trust in science to the principle of exclusion". (Darwin 1887 cited in Barlow 1958, p. 84.) Sadly, the principle of exclusion became quite popular among his followers.

Several modern Darwinians admitted the value of the concepts of directed evolution and even partly agreed with some of them. This could result in opposing statements on orthogenesis within the same work, even in Julian Huxley's major book on evolutionary synthesis. To create a unified view on evolution, Huxley tried to generalise all the available material. However, Darwinism, which Huxley believed in since childhood and which was a sort of religion for him, was the only theory used as a basis for all the generalisation.

A dog in a story by Franz Kafka remarks that science "has enough elasticity to admit exceptions and isolated cases" (Kafka 1971). This phrase seems an appropriate characteristics of Huxley's synthesis. Concerning orthogenesis, he wrote "The trends seen in adaptive radiation would appear to present no difficulties to the selectionist, and it is hard to understand why they have been adduced as proof of non-adaptive and internally-determined orthogenesis." (Huxley 1944, p. 494). Later in the same text, however, Huxley does not find it hard to understand at all: "Besides the usual trends constituting the radiation of a group, most of which, as we have seen, appear clearly to be towards adaptive specialization, there are others for which no adaptive significance has as yet been found." (Ibid.) From this discussion one may conclude that something similar to orthogenesis exists but is insignificant. "Orthogenesis of this sort, playing the major part in guiding evolutionary change, with selection in a purely limiting and subsidiary role, may be called dominant or primary. While dominant orthogenesis, if it exists at all, is rare and exceptional, what we may call subsidiary or secondary orthogenesis is common enough. Under the head of subsidiary orthogenesis I include phenomena which in the first place are of an orthogenetic nature in that they limit the freedom of variation and therefore of evolutionary change, and in the second place are subsidiary in that they merely provide limits within which natural selection still plays the main guiding and shaping role." (Huxley 1944, p. 504, 510).

Huxley speculated in a similar manner about adaptation or absence thereof and other phenomena. He thought that evolution had numerous ways, aspects, factors etc. They do not contradict the theory of natural selection but are merely unexplored yet. Something similar to orthogenesis may well be a "special case". This means that the possibility of orthogenesis was admitted even by an architect of the evolutionary synthesis.

Another modern Darwinist, **Ivan I. Schmalhausen**, was less indulgent to orthogenesis/autogenesis. "All these theories appeal to factors lying outside the scientific knowledge, and are hence unscientific. Instead of explaining, they introduce some obscure "principles" of development, which cannot be analysed. This explanation is not even erroneous but false." (Schmalhausen 1969, p. 145). He criticised severely all deviations from the natural selection theory, but after scathing criticism usually conceded that "developmental tendencies", "a common basis" etc. were important in evolution. His brief comments on these "tendencies" occupy an infinitesimal part of his writings, with the lion's share being given to the analysis of natural selection.

A similar analysis of directionality was presented in a number of works on parallelisms and regularities in the fossil record (Gilyarov 1970, 1985; Vorobyeva 1980). Their introductions contain critical comments about an "idealistic" interpretation of evolution but the main texts present facts and ideas agreeing well with orthogenesis. The authors noted that parallelisms and convergences could not always be explained by a similar selection pressure and that their adaptive significance was not always clear. Various lineages of lobe-finned fishes demonstrate parallelisms in the continuous change of skin dents in the tectorial bones of skull, in the evolution of stegocephal jaws, in the replacement of a rhomb scale by a round one, in change of rugosity of the dents etc. (Vorobyeva 1980). These phenomena were explained by a "historically formed organisation", "tendencies", "the role of initial organisation", "a common basis referred to by Schmalhausen", "the dominating role of the organisms nature stressed by Darwin" etc. The authors grouped these phenomena under the heading of "factors determining independent similarities", noting that they were still insufficiently studied. At the same time, they noted that one could not ignore regularities and directionality: "Phylogeny is a regular process having a definite direction. The question of directionality has recently been ignored in our literature. Idealistic interpretations of directionality presented by the supporters of orthogenesis (Eimer) or similar concepts (Berg, 1922; Osborn, 1930 etc.) resulted in denying or ignoring the obvious facts of directionality of evolution of several groups." (Gilyarov 1970, p. 9.). All this sounds ambiguous. Orthogenesis is accepted but, as it were, in inverted commas though the differences between orthogenesis and "orthogenesis", if any, are hard to grasp. It was either as a masked support of orthogenesis or else dogmatism and persistent reluctance to accept anything contradicting the natural selection theory.

Gilyarov also suggested that the directionality of evolution was determined by the "objectives" set for the organism by the environment. "The process of adaptation to new conditions during phylogenesis of different groups of land arthropods can evidently be considered as a process characterised by a definite direction, i. e. an orthogenetic process. However, there is no mystics in such an orthogenesis. The evolution is directed by the general principles of the phylum's organisation, which have evolved in the previous (aquatic) environment and by the principles of the most effective possible changes of this organisation in the new environment (on land)." (Gilyarov 1970, p. 238). The directions are determined by the ways of solving functional objectives evolved during the organism's adaptation to the environment; these ways are few and sometimes unique. For instance, active movement requires bilateral symmetry, while sedentary life does not. At the same time, there are "features of the initial organisation" determining the possible ways of adaptation. They also result in parallelisms. Parallelisms such as an independent origin of tracheal system in various groups during transition from life in water to life on land (arachnids, insects, myriapods) are "striking"; at the same time, they are "a common and stable phenomenon". Numerous convergences emerge, and the direction of evolution becomes an automatically regulated process, whose stages are governed by complex feedbacks (Ibid.)

Yuri V. Mamkaev presented a similar analysis of stinging organoids of protozoa. "Once a certain biological objective is set, it is solved in all ways possible on the given morphological base. All mechanisms appear which can be offered by the present organization" (Mamkaev 1987, p. 199). Some groups demonstrate all or most theoretically possible types, that is, the entire morphological spectrum, while others demonstrate a single type but with numerous variations on the topic.

It would seem that the analysis of directionality in relation with functional objectives might solve the problem of the inclusion of orthogenesis into Darwinism but in fact the advocates of orthogenesis were resolutely opposed to teleology, that is, the claim about the existence of purpose in evolution and development. In the texts cited above evolving organisms are like soldiers: the headquarters issue them orders even if there is no need for action, just to keep them occupied. It would seem that "somebody" always forces organisms to adapt to a new environment even if they are already adapted quite well. Some organisms are forced to be sedentary, while the others are forced to move actively. That is why the former acquired radial symmetry and the latter, bilateral symmetry. "Somebody" forced arthropods to leave the aquatic environment, which necessitated changes of the respiratory system, the covers etc. These changes, in turn, entailed modifications of other organs etc. Such speculations often provoke criticism of Darwinism. It seems logical that changes appear first and are put to some use afterwards. In the above examples, however, organisms seem to transform rationally in order to solve an objective.

Evidently, the authors of the texts cited above would brush off this criticism saying that their seemingly teleological speculations are just a metaphor. Nobody put any objectives to arthropods. They varied in numerous directions and those favoured by the environment were realised. However, it is difficult to distinguish metaphors from "usual" text in this case. Moreover, there is an evident tautology there: objectives set to organisms are judged upon by the organisms' structure, while the structure is judged upon based on the objectives set by a changing environment. Even if we mentally "edit" the texts removing all teleological or seemingly teleological elements, an uncertainly will remain. What exactly are "tendencies", "material for selection", "predisposition to vary" etc.? This mysterious side of the evolutionary process inevitably raises the spectre of orthogenesis.

Questions like these re-emerge continuously. It is considered that the most important change of the modern Darwinism was the incorporation of the developmental aspect. The traditional scheme "selection-mutation" contains a gap—the distance between the mutation and the phenotype exposed to selection. Developmental biology fills this gap revealing the forces influencing the path from gene to phenotype. One of them is the constraint on the number of possible directions of variation. This seems like a little indulgence for orthogenesis. However, modern authors analysing constraints keep returning to the Darwinian view on variation (it is discussed below in the chapter devoted to constraints on variation).

References

Boesiger E (1980) Evolutionary biology in France at the time of the evolutionary synthesis. In: Mayr E, Provine W (eds) The evolutionary synthesis: perspectives on the unification of biology. Harvard Univ Press, Cambridge (MA)/London, pp 309–322

Bowler PJ (2003) Introduction (Predislovie). In: In the shadow of Darwinism. Alternative theories of evolution in the 20th century (V teni darvinizma. Al'ternativnye teorii evolyutsii v XX veke). Yasnyy den', St Petersburg, p 5–6

Cain J (1993) Common problems and cooperative solutions. Organizational activities in evolutionary studies, 1937–46. Isis 84(75):1–25

Cain J (1994) Ernst Mayr as a community architect: launching the society for the study of evolution and the journal evolution. Biol Philos 9(3):387–429

Cuénot L (1951) Biological evolution. The facts. The uncertainties. (L'évolution biologique. Les faites. Les incertitudes). Avec la collaboration de A. Tétry. Masson et C-ie, Paris

Darwin Ch (1887) cited in Barlow N (ed) (1958) The autobiography of Charles Darwin 1809–1882. With the original omissions restored. Edited and with appendix and notes by his granddaughter Nora Barlow. Collins, London

Devillers C, Chaline J (1989) The theory of evolution. State of the question in the light of current scientific knowledge. (La théorie de l'évolution. Etat de la question à la lumière des connaissances scientifiques actuelles). Dunod, Paris

Dobzhansky Th (1937) Genetics and the origin of species. Columbia Univ Press, New York

Gall YaM (1988) Development of the natural selection theory: ecologic-genetic synthesis (Razvitie teorii estesvennogo otbora: ecologo-genetichesky syntez). Doctor's thesis. LGU, Leningrad

Gilyarov MS (1970) Regularities of adaptations of arthropods to life on land (Zakonomernosti prisposobleniy chlenistonogikh k zhizni na sushe). Nauka, Moscow

Gilyarov MS (1985) Parallel directions of evolution of insects and higher vertebrates (Parallel'nye napravleniya evolyutsii nasekomykh i vysshikh pozvonochnykh). In: Morphological studies of animals (Morfologicheskie issledovaniya zhivotnykh). AN SSSR, Moscow, pp 61–77

Gould SJ (1977) The return of hopeful monsters. Nat Hist 86:22–30

Gould SJ (1982) Darwinism and the expansion of evolutionary theory. Science 216:380–387

Grassé P-P (1962) The uncertainties of evolution (Les incertitudes de l'évolution). L'année biologique, 4 Serie, 1(7–8):375–389

Guyénot E (1961) The origin of species (L'origine des espèces). Presses universitaires de France, Paris

Heberer G (ed) (1943) The evolution of organisms (Die Evolution der Organismen). Gustav Fischer, Jena

Hossfeld U (2003) Otto Kleinschmidt (1870–1954) and "The scientific inadequacy of Darwin's work about the origin of species" (Otto Klyaynshmidt [1870–1954] i «Nauchnaya nepolnotsennost' raboty Darvina o proiskhozhdenii vidov»). (In: Evolutionary biology: history and theory (Evolyutsionnaya biologiya: istoriya i teoriya). SPbFIIET RAN, St Petersburg, pp 189–199

Hull D (1988) Science as a process. Univ of Chicago press, Chicago

Huxley JS (ed) (1940 [1978]) New Systematics. Reprint of the 1940 ed. Oxford University Press, Oxford

Huxley JS (1944) Evolution. In: The modern synthesis, 3rd edn. George Allen and Unwin, London

Jeannel R (1950) The march of evolution (La marche de l'évolution). Edition de muséum, Paris

Kafka F (1971) English edition: Investigations of a dog (trans: Muir W, Muir E.). In: Kafka F. The complete stories. Schocken Books Inc., New York

Khozatsky LI (1973) On several aspects of evolution (O nekotorykh storonakh napravlennosti evolyutsii). In: Some philosophical questions of modern natural sciences (Nekotorye filosofskie voprosy sovremennogo estestvoznaniya), LGU, Leningrad, 1:109–122

Khozatsky LI (1981) On directionality of evolution of land vertebrates (O napravlennosti evolyutsii nazemnykh pozvonochnykh). In: Life on ancient continents, its formation and development (Zhizn' na drevnikh kontinentakh, ee stanovlenie i razvitie). Nauka, Leningrad, pp 154–162
Kimura M (1983) The neutral theory of molecular evolution. Cambridge University Press, Cambridge
Laland K, Uller T, Feldman M, Sterelny K, Müller GB, Moczek A, Jablonka E, Odling-Smee J, Wray GA, Hoekstra HE, Futuyma DJ, Lenski RE, Mackay TFC, Schluter D, Strassmann JE (2014) Does evolutionary theory need a rethink? Nature 514(7521):161–164
Lukin EI (1940) Darwinism and geographical regularities in the changes of organisms (Darvinizm i geograficheskie zakonomernosti v izmenenii organizmov). AN SSSR, Moscow, Leningrad
Mamkaev YuV (1987) The problem of parallelisms and morphological radiation (Problema parallelizmov i morfologicheskaya radiatsiya). In: Polyansky YuI (ed) Evolutionary ideas in histology and embryology. (On occasion of the centennary of birth of academician Alexei Alexeevich Zavarzin) (Evolyutsionnye idei v gistologii i embriologii. [K 100-letiyu so dnya rozhdeniya akademika Alekseya Alekseevicha Zavarzina]). Trudy Leningradskogo obshchestva estestvoispytatelei, 86(1):195–203
Mayr E (1942) Systematics and the origin of species from a viewpoint of zoologist. Harvard University Press, Cambridge, MA
Mayr E (1963) Animal species and evolution. Belknap Press of Harvard University Press, Cambridge, MA
Mayr E (1970) Populations, species, and evolution. Belknap Press of Harvard University Press, Cambridge, MA
Mayr E (1982) The growth of biological thought. Harvard Univ Press, Cambridge, MA
Mayr E, Provine W (1980) The evolutionary synthesis: perspectives on the unification of biology. Harvard Univ Press, Cambridge, MA/London
Müller GB (2007) Evo-devo: extending the evolutionary synthesis. Nat Rev Genet 8:943–949
Nazarov VI (1981) The significance of studies of G. Tessier for the formation of the synthetic theory of evolution (Znachenie issledovaniy Zh. Tess'e dlya formirovaniya sinteticheskoy teorii evolyutsii). In: The problems of the contemporary history of the evolutionary teaching (Problemy noveyshey istorii evolyutsionnogo ucheniya). Nauka, Leningrad, pp 136–148
Nazarov VI (1991) The teaching about macroevolution. On the ways towards a new synthesis (Uchenie o makroevolyutsii. Na putyakh k novomu sintezu). Nauka, Moscow
Olsson L (2003) Resistance to Darwinism in Sweden: idealistic morphology, Lamarckism, the idea of constancy of species and antiselectionism (Soprotivlenie darvinizmu v Shvetsii: idealisticheskaya morfologiya, lamarkizm, ideya postoyanstva vidov i antiselektsionizm). In: In the shadow of Darwinism. Alternative theories of evolution in the 20th century (V teni darvinizma. Al'ternativnye teorii evolyutsii v XX veke). Yasnyy den', St Petersburg, pp 82–94
Pigliucci M (2007) Do we need an extended evolutionary synthesis? Evolution 61(12):2743–2749
Provine WB (1987) Progress in evolution and meaning of life. In: Wakers CK, Van Helden A (eds) Julian Huxley. Biologist and statesman of science. Rice Univ Press, Houston, pp 165–181
Reif W-E (1983) Evolutionary theory in German paleontology. In: Grene M (ed) Dimensions of darwinism. Cambridge Univ Press, Cambridge, pp 173–203
Rensch B (1943) Palaeontological evolutionary rules from zoological viewpoint (Die palaontologischen Evolutionsregeln in zoologischer Betrachtung). Biologia generalis 17(1–2):1–56
Rensch B (1954) New problems of the evolutionary theory: transspecies evolution (Neuere Probleme der Abstammungslehre: transspeciphische Evolution). Enke, Stuttgart
Rostand J (1962) The past of transformism (Le passé du transformisme). L'année biologique, 4 Serie, 1(7–8):339–359
Schmalhausen II (1938) The organism as a whole in the individual and the historical development (Organizm kak tseloe v individual'nom i istoricheskom razvitii). AN SSSR, Moscow-Lenngrad
Schmalhausen II (1939) Pathways and regularities of the evolutionary process (Puti i zakonomernosti evolyutsionnogo protsessa). AN SSSR, Moscow
Schmalhausen II (1946) Factors of evolution (Faktory evolyutsii). AN SSSR, Moscow-Lenngrad

Schmalhausen II (1969) The problems of Darwinism (Problemy darvinisma). Nauka, Leningrad
Severtsov AS (1990) Directionality of evolution (Napravlennost' evolyutsii). MGU, Moscow
Simpson GG (1944) Tempo and mode in evolution. Columbia Univ Press, New York
Stahl FW (1988) News and views: a unicorn in the garden. Nature 355:112–113
Stebbins GL (1950) Variation and evolution in plants. Columbia Univ Press, New York
Stebbins GL (1980) Botany and the synthetic theory of evolution. In: Mayr E, Provine W (eds) The evolutionary synthesis: perspectives on the unification of biology. Harvard Univ Press, Cambridge, MA/London
Teissier G (1962) Modern transformism (Transformisme d'aujourde'hui). L'année biologique, 4 Serie 1(7–8):359–375
Timofeeff-Ressovsky NW (1939) Genetic and evolution (a report of a zoologist) (Genetik und evolution [Bericht eines Zoologen]). Z Indukt Abstamm Vererbungsl 76:158–219
Towards a new synthesis in evolutionary biology (1987) Circular of the fifth international meeting on evolutionary biology organized by the Laboratory of evolutionary biology, Czechoslovak Academy of Sciences. Prague, Czechoslovakia, 5–11 July 1987
van Waesberge H (1982) Towards an alternative evolution model. Acta Biotheor 31:3–28
Vandel A (1954) Evolution considered as a phenomenon of development (L'évolution considérée comme phénomene de developpement). Bull Soc Zool Fr 79(4):341–356
Vandel A (1958) Man and evolution (L'homme et l'évolution). Gallimard, Paris
Vandel A (1964) Biospeleology. The biology of cave animals (Biospéologie. La biologie des Animaux Cavernicoles). Gauthier-Villars Éditeur, Paris. English edition: Vandel A (1965) Biospeleology. The biology of cavernicolous animals. Pergamon Press, Oxford
Vandel A (1968) The origin of life (La genèse du vivant). Masson et C-ie, Paris
Vorobyeva EI (1980) Parallelisms and convergencies in the evolution of crossopterygian fishes. Morphological aspects of evolution (Parallelizmy i konvergentsii v evolyutsii kisteperykh ryb. Morfologicheskie aspekty evolyutsii). Nauka, Moscow, pp 7–29
Vorontsov NN (1980) Synthetic theory of evolution: its sources, main postulates and unsolved problems (Sinteticheskaya teoriya evolyutsii: ee istochniki, osnovnye postulaty i nereshennye problemy). Zhurnal vsesoyuznogo khimicheskogo obshchestva im. D I Mendeleeva, 15(3):295–316
Wintrebert P (1952) Position of chemical Lamarckism in the determinism of evolution (Position du lamarckisme chimique dans le déterminisme de l'évolution). Vie et milieu 3(3):227–231
Wintrebert P (1962) Living world, the creator of its evolution (Le vivant, créateur de son évolution). Masson et C-ie, Paris

Chapter 6
New Concepts of Directed Evolution in the Recent History of Evolutionary Biology

Abstract In the recent history of evolutionary biology new concepts of directed evolution have been suggested in various countries: typostrophism (Germany), organicism (France), *shutayzey*/subjectivity (Japan), Frankfurt theory (Germany), contours of evolution (Russia), autoevolution (Sweden) and biological synthesis (Venezuela). The amount of data on directionality increased but new terms and seemingly new ideas confused the field further. In this way, the history of the idea of directed evolution repeated itself.

Keywords Typostrophism · Organicism · *Shutayzey*/subjectivity · Frankfurt theory · Contours of evolution · Autoevolution · Biological synthesis

During the time when the modern evolutionary synthesis was being formed, some scientists elaborated their own evolutionary concepts. A German palaeontologist **Otto Heinrich Schindewolf** (1896–1971) was one of them. He began to write on evolution as early as in the 1930s and developed his own doctrine (*Lehre*) called typostrophism (*Typostrophenlehre*) by the late 1950s. Schindewolf criticised both Lamarckism and Darwinism. From his viewpoint, these two "systems of doctrines" (*Lehrgebäude*) were really close to each other as they both looked for the source of evolutionary changes in the use or disuse of organs and the action of environment. Schindewolf thought that evolutionary mechanisms were internal. He focused on the gaps in the fossil record and the origin of "types". According to Schindewolf, drastic evolutionary changes described by palaeontologists could not be explained by an incompleteness of the fossil record, as Darwinists claimed. Similarly to many other evolutionists, he considered the evolution of horses as a vivid example. It is studied in much detail and there are both gradual transitions among the horses' ancestors and several gaps between genera (*Tetraclaenodon, Eohippus*). These genera seem to have originated not just very rapidly but explosively, without any intermediate forms, after which they started to evolve gradually. Schindewolf found similar examples in the evolution of corals, ammonites and other organisms. He noted similarities of final evolutionary stages with abnormal forms of the initial ones, interpreting this circumstance as an indication of the unity of evolutionary and developmental laws. Similarly to other proponents of directed evolution, Schindewolf

I. Popov, *Orthogenesis versus Darwinism*,
https://doi.org/10.1007/978-3-319-95144-7_6

pointed to various parallel rows. In general, however, he was more interested in drastic changes. He left open the question about the mechanisms of the origin of "types", making only a few suggestions such as that it was in some way similar to the emergence of abnormalities and modifications under the impact of biologically active substances.

Schindewolf considered the origin of species as a question of minor importance in evolutionary biology. "Speciation proceeds at the periphery of the definite way of evolution, providing, as it were, only external ornaments and arabesques." (Schindewolf 1947, p. 12). His own concept dealt with "types". Schindewolf used this term in a broad sense: a type was a taxon demonstrating sharp differences from others. For instance, he considered all horse-like mammals as belonging the same type but could also apply the name "type" to several of their genera.

Typostrophism can be summarised as follows. The evolution of any group of a high taxonomic level consists of three phases: typogenesis, typostasis and typolysis. Typogenesis is a stage of explosive evolution. Natural selection is uninvolved into it because it requires longer time periods and also because the characters of types are neutral. The second stage, typostasis, is a gradual evolution of the type demonstrating numerous intermediate variants. Typostasis is a revelation of pre-existing possibilities rather than the creation of anything new. Types change not by addition of species characters up to the level of genera, family etc. but by direct transition of type characters from one class to the other. Orthogenesis occurs at the stage of typostasis, while the role of natural selection is insignificant. Typostasis is followed by typolysis, i. e. senescence and natural death of a taxon (Schindewolf 1936, 1993). Schindewolf claimed that typogenesis provided an insight not only into the evolutionary process but also into the hierarchy of organisms, which is could not be understood in terms of Darwinism at all.

It can be easily seen that typostrophism is similar to, not to say identical with, the main ideas of Dmitry N. Sobolev, Leo S. Berg and Daniele Rosa or even those of Alpheus Hyatt, who was active 80 years before the emergence of typostrophism. Nevertheless, it was presented as a new doctrine. The typostrophic theory of evolution had its followers, and Schindewolf's terms were used quite frequently. It has even been suggested that typostrophism dominated among German palaeontologists over the following decades (Reif 1983).

An idea of organicism is close to typostrophism. It also focuses on the similarity of life cycles of the taxon and the individual. A French zoologist **Albert Vandel** (1894–1980) was actively developing this theory approximately at the same time when Schindewolf was developing his. Vandel conducted numerous studies, spanning almost 60 years, on several groups of invertebrates such as flatworms, ants and woodlice. He was especially interested in animals living in caves and other underground habitats. In 1948 he organised a special laboratory for the study of such animals and remained its head for 18 years. Vandel was interested in biogeography, drift of continents, general problems of taxonomy and evolution as well as philosophy (Angelier 1982).

The main thesis of Vandel's concept was that "All life, both individuals and phyletic lines, is fated to perish. As the phyletic lines age, the animal organisation degrades and thus all the zoological groups are led ineluctably towards their end and disappearance." (Vandel 1965, p. 463). Similarly to organisms, phyletic trends originate, develop, flourish, age and die. The first stage of their evolution, i. e. typogenesis (Vandel cited Schindewolf and used his terminology), was obscure. Vandel did not discuss it in detail and suggested only a few ideas about "mechanisms of phyletic rejuvenation". He considered it self-evident that that new types originated from those that, though not necessarily aged, had at least undergone some evolution. In order for the old types to transform into new, they must be rejuvenated. Following **Walter Garstang** (1868–1949), Vandel supposed that "pedomorphosis", i. e. maturation at the "wrong" stage of individual development, was one of such mechanisms. Such cases were usually considered as exceptions but Vandel thought that they were a general phenomenon.

Once originated, a new type evidently evolves, and does so "orthogenetically". Vandel supported the idea of orthogenesis and thought that organisms varied spontaneously in definite directions irrespective of the changes in the environment. One of his arguments was that organisms evolved even in a constant environment (e.g. in caves). At the same time, Vandel emphasised that evolution was prolonged and continuous. Analysing zoological data, he demonstrated gradual patterns of changes of the organisms' structure and of transition from one habitat to the other.

Organicism was elaborated based on the data of biospeleology, the discipline studying organisms living in caves and other underground shelters. (Vandel was said to know about cave dwellers more than anyone else.) Vandel believed that cave animals were especially interesting for evolutionary studies because they represented the final stages of the life of taxa. Vandel discussed adaptation and natural selection but, unusually, did so outside the context of evolutionary mechanisms. He criticised Darwinism and especially the idea about the absence of non-adaptive characters. He thought that Darwin's concept summarised an ancient teleological thesis proclaiming a universal character of finalism: all that was unnecessary or non-adaptive in an organism was automatically eliminated. Vandel was resolutely opposed to finalism and teleology and claimed that non-adaptive changes were common in evolution.

According to Vandel, the capacity for adaptation is high during the stage of flourishing and specialisation of a group. Then it decreases, and the group evolves numerous non-adaptive characters, much in the same way as an aging organism becomes more prone to diseases. At the final stages of evolution, degenerative features such as the loss of sight, depigmentation etc. become more important, so Vandel used the term adaptation in inverted commas in this case. These characters might seem adaptive but to consider them as such would be the same as to consider catarrh, rheumatism and other age-related diseases as an adaptation to senescence.

Vandel also speculated about "autoregulation"—the capacity of organisms to adapt, i.e. to react purposefully to external conditions. He did not discuss its mechanisms, taking the existence of this phenomenon as an axiom. A decreasing capacity to autoregulation may result in "ultra-specialised forms" adapted to very specific conditions. A remarkable example is provided, again, by caves, which become "a prison before becoming a grave" for ultra-specialised animals that happened to occur there. In general, however, Vandel thought that adaptation was an extremely complex question. If an organism has a poor sight, this can be compensated by organs of touch, echolocation etc. and such an organism can be considered as harmonious and adapted. Vandel noted that things were not so simple because negative cases were at least as common, if not more so, as positive ones: a decrease of some "adaptive" capacities was sometimes uncompensated for by any others. Non-adapted organisms do occur.

The question of competition was equally complex. Vandel pointed out that it was difficult to analyse the role of competition in evolution because different scientists understood this word differently. Vandel considered it as a competition between species and gave several examples but did not seem to think them important for understanding evolution.

Finally, Vandel analysed preadaptation. In general, he was in favour of this idea. He emphasised the fact that cave dwellers underwent evolutionary changes not because they had settled in caves but, on the contrary, they populated caves because they had acquired features making life under "normal" conditions impossible. Vandel believed that the origin of cave animals was a kind of an accident that occurred in different lineages in different periods. Gigantism probably originated in the same manner, as a consequence of the taxon's senescence. However, Vandel thought that the term "preadaptation" was inappropriate for the final stages of the evolution of the taxon. A "period of preparation" would be more exact because "regressions and reductions" prevailed over adaptation. During this time animals evolve nutritive and ecological "preadaptations" (or regressions): morphological changes (albinism, micro- and anoftalmia and apterism), physiological changes (low metabolism, hypofunction of endocrine glands, slow development, loss of pigmentation), sensory and ethological changes (simplification of behaviour, weakening of trophisms, search for humid conditions, inclination to burrow and search for food relying on the sense of touch and smell rather than sight). Such "preadaptation" results in the origin of animals that have to hide in dark and wet habitats because elsewhere they would be exterminated by "normal" competitors under "normal" conditions. Vandel came to the conclusion that caves had always been a refuge for animals, as they were for humans.

The concept of organicism is somewhat incomplete. It contains contradictions and ignores some problems such as adaptation or autoregulation. However, one cannot deny it originality and wit. Vandel applied interesting approaches to the analysis of evolution and it would be difficult to disprove his major points.

Vandel was a friend of a French philosopher and palaeontologist **Pierre Teilhard de Chardin** (1881–1955) (Angelier 1982), who has also been named as an advocate of orthogenesis. His major treatise, the famous *Le phénomène humain* (*The Phenomenon of Man*), was published in the 1950s, when the modern evolutionary synthesis was being established in biology. Teilhard de Chardin discussed orthogenesis in this work (Teilhard de Chardin 1987) but his views on this topic were quite complex for personal reasons.

Since his youth Teilhard de Chardin was a member of Jesuit Order, which did not approve of evolutionary studies. His contacts with the Order did not cease throughout his scientific career. He studied palaeontology but had difficulties with the presentation of his results and ideas. He was especially interested in the evolution of man and tried to bring scientific concepts in agreement with theology, that is, to reconcile incompatible things. These travails influenced his theories. In the introduction to *The Phenomenon of Man* Teilhard de Chardin noted that he wanted the readers to consider his work as purely scientific (Teilhard de Chardin 1987) but this was hardly possible. *The Phenomenon of Man* is a philosophic assessment of the origin of man, which is seen a directed, i. e. orthogenetic process. Teilhard de Chardin did not discuss evolutionary tendencies or criticise Darwinism in this treatise though he had some material for this as indicated by his studies of extinct rodents *Siphneidae*. He showed that these rodents evolved in three parallel trends, the parallelism being reflected in tendencies towards increasing size, merging of neck vertebrate and hipsodontia. The main conclusion from these studies was that: "syphneids demonstrated directed morphological transformation, evolution, so called orthogenesis, it is impossible to deny this phenomenon" (Teilhard de Chardin 1950, p. 173). Teilhard de Chardin emphasised that the most significant orthogenetic direction was the one that had resulted in the origin of man.

The next notable person in the history of orthogenesis was a French-Italian-American scientist **Léon Camille Marius Croizat** (1894–1982). His scientific carrier was also complicated. Croizat was interested in biology since childhood and, having grown up in well-to-do family, obtained a good knowledge of biology while still a schoolboy. In particular, he became familiar with the studies of the zoologists from Turin, the followers of Daniele Rosa. Later, however, fortune turned away from him. Because of financial difficulties Croizat had to choose other occupations, and when the fascists came to power in Italy he emigrated to America, where he earned his living by chance jobs. Finally, he became a member of a group identifying plants in the parks of New York and then obtained a temporary position in the Arnold Arboretum in Harvard.

Croizat was especially interested in succulent plants. At the same time, he tended to broad generalizations in botany and biology. He decided to study all significant botanical works published over last 300 years. Croizat was well equipped for this task because he knew several foreign languages well ("I have a free knowledge of French, Italian, Spanish, Portuguese, Latin, Russian, German, and, with a dictionary, read Greek", he wrote (cited in Panke and Danova 2003, p. 98). He probably carried out much of what he had planned. He read many other works revealing deeper sources of modern views. In particular, he found that the question about the centres of origin of groups of organisms, which especially interested him, was first asked by Spanish priests who, soon after discoveries of Columbus, tried to find out where American Indians came from (Croizat et al. 1974, p. 273).

Croizat showed great erudition and enthusiasm but his works cannot really be considered scientific as he ignored the basic principles of scientific studies. Croizat wrote whatever he wanted, without paying the slightest attention to comments or objections of his colleagues. As a result, he became an outsider of the scientific community. His books were mostly published at his own expense and he had difficulties in finding a job. Finally, Croizat achieved success in Venezuela, where he emigrated in 1947 when he was 52. There, far from large scientific centres, he became a "professor" and could do whatever research he felt like doing (Panke and Danova 2003).

In the 1950–60s Croizat published several voluminous treatises on various biological problems of biology. At first he edited three volumes of *Panbiogeography* (Croizat 1958), more than 1500 pages each. Then he prepared their brief version (about 2000 pages printed in small print) entitled *Space, Time and Form: Biological Synthesis* (Croizat 1962). In these treatises he presented a review of classification and geographical distribution of the living organisms from a viewpoint reminiscent of ologenesis. Croizat dedicated his "biological synthesis" to Daniele Rosa, considering ologenesis and orthogenesis as its main principle. Other principles of the new synthesis were: evolution = space + time + form; geographic distribution is not the same as dispersion; dispersion is always secondary; adaptation is based only on "orthogeny"; variants of morphological structures result from a general plan; Earth and life evolved together; distribution patterns of fossils and living organisms are united. Croizat did not doubt that his alternative evolutionary synthesis would sooner or later triumph in biology (Croizat 1962).

Generalising everything, Croizat allowed himself extensive digressions. He could suddenly lapse into colloquial language or start a long and mocking enumeration of errors and failures of his opponents. Apparently, most of his colleagues simply did not read what he wrote. In general, the outcome of Croizat's studies seems meagre in comparison with his efforts. They did, however, have some resonance in the scientific community. He had several followers who took the trouble to bring his ideas in correspondence with the principles of scientific research. In the end, however, Croizat quarrelled with them as well. Nowadays it is quite impossible to get a complete idea of his "biological synthesis": no one would read all these books and the commentaries of his followers are incomplete and inexact.

Biogeographic observations were probably a major source of *Biological Synthesis*. Croizat painstakingly marked the habitats of related groups on maps connecting them with lines, which he called "tracks". Croizat believed that a track was a chain of centres of origin of new forms; each centre appeared due to the separation of the habitat of a previous form of lower taxonomic level (Croizat 1960 cited in Panke and Danova 2003). These tracks were not confined to one continent but joined areas separated by various climatic and physical barriers such as South America, South Africa, Madagascar and India. Croizat claimed that this was because everything on Earth, living as well as non-living, had a common history. Similarly to Daniele Rosa, he believed that a common ancestor inhabited the entire area limited by tracks, then it evolved in the same manner everywhere in its distribution area even if it became divided into isolated parts. Surprisingly, he denied the theory of continental drift even though it agreed with his ideas. Some followers of Croizat pointed that he found new facts in support of orthogenesis (Grehan and Ainsworth 1985): if previously orthogenesis was supported by studies of "form", Croizat found biogeographic evidence, the facts of parallel evolution of geographically isolated forms. However, similar facts were cited before him by Vandel, Rosa and other authors. The treatises of Croizat are striking but after a close study they give the same impression as other "heresies" in evolutionary biology: what is new in them is untrue, what is true is not new.

Strangely, another "new" concept reminiscent of orthogenesis appeared in Germany. It is known as the Frankfurt constructional theory as well as by several other names. (The authors seem to see no difference between the notions of "theory", "doctrine", "concept", "paradigm" and "school"). The German tradition of evolutionary studies including orthogenesis is long and rich but the Frankfurt authors chose to ignore their predecessors and proclaimed a new version of directed evolution. Their papers usually start with the statement that new (and correct) views on evolution and morphology have been developing in the Senckenberg Research Institute since 1970s. Darwinism is referred to as an outdated theory, which has been disproved by the Frankfurt authors and is gradually dying off (Edlinger et al. 1991; Gutmann and Weingarten 1992, 1994; Gutmann 1994, 1996, 1997 and others).

The main idea of the Frankfurt thinkers is as follows: evolution is an internally directed process, which could not have occurred otherwise. All its possible ways are being realised. One of the articles presents a speculation on the colonisation of the aquatic environment by tetrapods. Having collected numerous examples of swimming vertebrates and illustrated them with very good pictures, the authors came to a highly satisfactory conclusion that tetrapods had realised all possibilities in their aquatic evolution. This observation was considered as an argument in favour of the Frankfurt theory (Gutmann 1994).

The Frankfurt authors did not bother with proofs of their viewpoint on evolution. They acted in the worst tradition of evolutionary thinking substituting a scientific concept with something close to a belief and explaining how one should consider biological facts in terms of their correct theory. One of their articles is devoted to preaching about the familiar terms selection and adaptation. The authors pointed out that "selection" and "adaptation" were not mentioned in the works of the Frankfurt school and decided that the readers needed an explanation. These terms were not mentioned because they were meaningless. It was illustrated by a schoolbook example of peppered moths (*Biston betularia*) demonstrating industrial melanism—the dominance of dark individuals in a polluted environment. From the viewpoint of Darwinism, dark moths are more numerous near industrial centres because light ones perish. The cause of their death are birds, which eating light moths. From the viewpoint of the Frankfurt theory, the cause is not birds but "construction". For the readers who do not understand "construction", an additional explanation was proposed: "organisation". The cause of the moths' death must be looked for in this construction/organisation, which can be realised by an organism. The death of light moths must be explained in the same way as that of fish taken out of water. Speculations about adaptation are inappropriate in this case because it is always an *post factum* analysis, which is always arbitrary and always correct (Grasshoff 1994).

The Frankfurt theory leaves a depressing aftertaste. It is a simplified version of the familiar criticism of Darwinism with an addition of a handful of new terms. There are few new arguments in favour of orthogenesis in their writings. Curiously, they sometimes repeat almost word-for-word some ideas of earlier authors, in particular, Leo Berg, though they are evidently not familiar with his works. So, Berg suggested that stereochemical properties of proteins determined directionality and the Frankfurt authors write that new data on molecular structures will provide information about the basis of internal evolutionary factors. Similarly to many advocates of orthogenesis, who claimed that their "laws" were neither mystical nor teleological, the Frankfurt authors consider that evolution is a directed but not a "purposefully directed" process (Gudo and Gutmann 2003, p. 185).

The Frankfurt theory is also similar to the previous versions of orthogenesis in the criticism that it invites. So, R. Kaplan noted that the Frankfurt authors simplified Darwinism claiming, for instance, that according to Darwin living organisms were exclusively the result of the action of the environment. Darwinism does not really mean that. The role of primary organisation is also significant, and Darwin admitted this. Therefore, there are no reasons for changing the paradigm (Kaplan 1992). Another typical critical remark concerning orthogenesis—that all its arguments can be easily explained in terms of Darwinism—was also repeated in connection with the Frankfurt theory (Remane 1983).

The Frankfurt theory can be probably best understood if we consider the following statement: "Evolution is an event for which there are no witnesses. To be more precise: the witnesses should be first called to the witness stand and questioned in such a way that they give us the desired information. In order to be able to understand and explain the modern fauna and flora as a result of evolution, we must describe and structuralise living organisms in such a way that they become accessible to a scientific treatment." (Gudo and Gutmann 2003, p. 174). If evolution cannot be observed, it naturally follows that one can fantasise, model or "structuralise" based on trivial facts or change paradigms to one's own liking.

In their approach to morphology, the Frankfurt authors got carried away by "structuralising", "modelling" and "constructing", the search for analogies and the composition of schemes of development, evolution and other biological processes. Based on such a methodology, they tried to determine the directions of evolution. The analogy between machines and organisms was seen an especially promising approach to "structuralising". Organisms were considered as "hydraulic constructions" or "transformers of energy" which could be described with the use of various characteristics of machines such as energy transformation, performance, perfection of construction etc.

It would seem that this approach is a revival of the old German *Naturphilosophie*, with its passion for analogy, and the polemics between mechanism and vitalism. However, the Frankfurt theory seems to have no historical roots. It is expressed in an entirely new language, hardly understandable for outsiders. New words are tirelessly coined and synonyms are piled to clarify ideas. Yet clarity is never achieved. If anything, the new language complicates the understanding of problems related to orthogenesis even further.

A Swedish cytogeneticist **Antonio Lima de Faria** also created a concept of the directed evolution from scratch. His style resembles that of the Frankfurt theorists. His main ideas are as follows: there are no random events in evolution, there is no natural selection, there is no theory of evolution (and there has never been one), all living and non-living matter evolves following the same laws. He referred to the unity of laws governing the structure of living and non-living matter as "periodicity". The books of Lima de Faria are copiously illustrated and written using a simple language without much effort at coherence. They contain a number of resolute claims. For example, the "periodicity" is characterised as follows: "hippopotami are related to pigs", "plants, fish, invertebrates and mammals have placentas", "insects, birds and mammals can fly", "a snowflake looks like a radiolarian" etc. Lima de Faria made long lists of such trivial facts and claimed that he was the first to pay attention to them. He dispenses with the question of the mechanism of periodicity with a remarkable ease: this phenomenon has just been discovered and its mechanisms are yet unknown (Lima de Faria 1995). To remind, in 1906 Wladimir M. Schimkewitsch published a paper *On the Periodicity in the Classification of Pantopods* (*Ueber die Periodizaet in dem System der Pantopoden*) (Schimkewitsch 1906). Periodical systems of biology were actively discussed in the 1920s, when Vavilov presented his law of homological series and Berg developed nomogenesis, as well as afterwards. However, Lima de Faria was apparently oblivious of the fact that "periodicity" in biology was discussed several times before him[1] (see Popov 2002).

Lima de Faria credited himself with the discovery of another important phenomenon, autoevolution. He remarked that he did not like to introduce new terms but was obliged to do so in this case. Now there is one more word in the long list of terms characterising directed evolution (autogenesis, orthogenesis, nomogenesis etc.) and readers have to spend even more time to untangle them. There are some differences between autoevolution and other orthogenetic concepts but they are not significant enough to introduce a new term and to claim priority. Lima de Faria suggested 56 "principles of autoevolutionism", which is many more than his predecessors did. There are among them some relatively "normal" ones such as no. 11: "All variation is ordered since it occurs within the frame allowed by the initial fixation. Order arises from order." (Lima de Faria 1988, p. 304). Others are quite weird such as no. 47: "A deer may be transformed into a whale by chemical manipulation and relatively rapid events. DNA cloning and hydridization shows that whales and seals do not derive from fishes but from land mammals such as deer and mink respectively. The seals use the same chemical mechanism against asphyxia, when they dive, that human fetuses use during prolonged birth" (Ibid., p. 309). It would seem that the author had believed that whales were derived from fishes and only new genetic data convinced him that this was not so. The discovery impressed him so much that he made it a "principle" of his theory. It is principles like this that are the new content that autoevolution has to offer autogenesis, orthogenesis, nomogenesis etc.

[1] Michael D. Golubovsky, an enthusiastic follower of Lyubishchev and Berg, had a conversation with Antonio Lima de Faria. Both were greatly surprised: Golubovsky by the fact that Lima de Faria did not know about Berg's treatises and Lima de Faria, by the fact that somebody had expressed similar ideas before him (Golubovsky, personal communication).

Lima de Faria stressed an analogy between living and non-living worlds more than other adherents of orthogenesis. In this way he returned to the pre-evolutionary stage of biology, when the study of minerals and living organisms was considered as the same field of research. It would seem that time has made a full circle and we should start making classifications where snowflakes, radiolarians and other radially symmetric natural objects would belong to the same taxon. It is indeed tempting to try to create such a beautiful system but no one has succeeded yet. The books of Lima de Faria are a good illustration of such an attempt and its failure. They have all the drawbacks of theoretical biological treatises such as bulky collections of facts, hasty generalisations, the lack of scientific argumentation and a style which is critical without being creative.

The idea about spontaneous evolution flourished in the opposite part of the world independently of the Frankfurt morphologists and Lima de Faria. Its source were the studies of monkey behaviour by a Japanese scientist Imanishi Kinji. In the 1970s he conducted a series of interesting observations on the development of intellectual capacities in monkeys living on an island and suggested some ideas concerning evolution. They were discussed and developed by several Japanese authors who organised the so-called Osaka-group together with several "structuralists" from other countries in 1986. This union was based on the denial of the leading role of natural selection in evolution (see Vorobyeva 1991).

A summary of "Japanese orthogenesis" can be seen in a recent theory of "subjectivity-self-organization" by **Kunio Kawamura** (Kawamura 2002; Kawamura 2003). This author emphasises the originality of his theory and its close connection with the Japanese scientific and cultural tradition. "One notable nature of the Japanese views on life is that people realize continuity of human, other living organisms, and several natural phenomena, in which they feel that human is governed by and involved in nature. People feel spirits not only in the living organisms such as just insects, but also in nature and natural phenomena such as mountain, river, ocean, the sun, and even natural disasters. Although this view may include some common sense as appeared in animism in many places of the world, it should be noted that the view is highly sophisticated as much as it could not be called as primitive. Facts that humans are capable to make life on the wealth of nature remind us that nature must be deeply respected. Thus, it is possible that Japanese scientists create different scientific theories from those created by scientists with different cultures (Kawamura 2003, p. 220).

Kawamura's concept is indeed original but its originality concerns form rather than content. There is basically nothing new in it. This is closely related with the fact that Kawamura knew nothing about the history of orthogenesis; in fact, he was probably uninterested in it because it belonged to the European culture. His concept was based on the works of Imanishi, and Kawamura sometimes did not even separate his own ideas from Imanishi's.

"The Japanese way of thinking" applied to evolutionary biology led to holism. A distinctive feature of the Kawamura's concept is an attempt to show the unity of the largest possible number of events associated with the evolution of biological objects, from the origin of life and to the evolution of human society. He tried to use as an argument the comparison of the processes involved in the origin of life and the formation of human cultures. According to him, culture and civilization of people are formed as a result of "self-organization", similar to that of organic molecules, which was postulated in the hypothesis of the origin of life by Oparin. At the same time, Kawamura argued that features well visible at the highest levels of the living matter organisation can also be found in the microcosm of small organisms. He is deeply convinced that bacteria have cultures similar to human ones. This interesting suggestion remained unexplained. The author only appealed to researchers urging them to find some non-genetic means of information storage in bacteria.

Kawamura stressed that he was not a biologist but a chemist. He thought that a "layman" could understand biological problems better because he was not influenced by stereotypes. "An advantage of the proposals from non-specialists of biology is that the ideas could capture perspective views rather than focus on microscopic views" (Kawamura 2003, p. 220). To remind, Darwin created his theory for 20 years, carefully collecting numerous facts and doing research in various fields of biology and geology. He often delved into the most minute details of particular problems. For example, he spent 10 years studying the taxonomy of barnacles, which that at first sight had no direct relation to the theory of natural selection. Darwin himself doubted whether his intensive studies of these creatures were reasonable. Nevertheless, such an approach certainly contributed to the solidity of Darwin's theoretical writings. The "broad view of a layman" looks poor in comparison.

However, the Japanese anti-Darwinists did not notice any incongruities and spoke condescendingly of Darwin's theory. They apparently consider their comments and Darwin's writings to be at the same level of scientific research. At the same time, the Japanese evolutionists generously admit that Darwinism was not all wrong. The concept of Imanishi and that of Darwin have things in common since the former does not completely deny the role of mutation and selection, while the latter did not consider natural selection as the only driving force of evolution (Kawamura 2003, p. 221). The conceit of contemporary authors seems to be progressing. Kawamura remarked that his concept was related to the works of Darwin and Imanishi in the same way as Euclidean geometry is related to non-Euclidean one (Ibid., pp. 235).

The "theory" of Kawamura and his supporters is a chaotic presentation of claims traditionally associated with the search for the weaknesses of Darwinism. Mentions of the similarity of ontogeny and phylogeny, the succession of plant communities, "bacterial cultures", endosymbiosis are scattered in a bizarre disorder. Similarly to the Frankfurt thinkers, the Japanese evolutionists do not feel the need to prove their point of view but present it as a teaching to be believed in. The "theory of subjectivity" is also written in its own language with many terms. At the same time, Kawamura noted that he did not attach much importance to terminology because there are things that the white man would not understand anyway such as the Japanese word *shutayzey*, which is the essence of understanding of the fundamental principle of life. The English word *subjectivity* is a mere shadow. In the subsequent translation into the Russian, the meaning of the concept was lost altogether. For a publication of the article in Russian it was translated as *individual'nost'* (individuality)

(Kawamura 2003) but the author was displeased about it. He wanted it to be translated as *subyektivnost'* but on being informed that it did not correspond to the English *subjectivity* he despaired and left everything to the publishers, whose patience was not unlimited. Evidently, translations into other languages would only increase misunderstanding. As a result, the central term of the greatest evolutionary theory remains obscure. A foreign reader must be content with the information that in order to understand evolution one has to enter into the spirit of the Japanese mind and grasp the existence of *shutayzey*/subjectivity/*individual'nost'* because it is the main characteristic of living matter.

Reading such a work, one imagines an eastern thinker surrounded by a circle of disciples reverently picking up every word of their teacher. Kawamura noted that Imanishi, his predecessor, spoiled the impression from his concept by "some non-scientific statements" such as "Organisms change to which they should be changed" (Kawamura 2003, p. 222). Such pronouncements can be made only if they remain unquestioned. This non-scientific approach to theoretical problems resulted in yet another parody of the evolutionary theory.

It seems that there were no such "new" concepts of orthogenesis in Russia. The support of directed evolution in Russia was usually associated, directly or indirectly, with the analysis of Berg's nomogenesis and the works of his followers. There was, however, one new concept in which the idea of directed evolution in central. It is recent or, rather, it recently came to light.

The works of **Eduard B. Akhnazarov** (1933–2001), which had started in the 1960s, were only published in 2002. His main work is called *Contours of evolution*. The title suggests that our knowledge of the evolution is limited and the researchers should just try to outline their ignorance. Akhnazarov believed that this was so. He wanted to create a framework of our knowledge but that had to be "the frame of the whole" (Akhnazarov 2002, p. 243), that is, the knowledge about the evolution not only of organisms but of everything in the biosphere, from "pre-life molecules" to the human society.

Akhnazarov was an economist, not a biologist, but regarded his work on evolution as his lifetime project. The amount of work he did impresses, and so does his dedication and self-sacrifice. For many years he worked without publishing anything. The paper *Contours of evolution* appeared as *samizdat* (self-published clandestine literature in the Soviet Union) in 1981. Akhnazarov worked on the new version for almost 20 years but could not complete it, leaving fragments and sketches to his followers. His work remained outside the academic evolutionary biology.

Despite its incompleteness and the lack of biological material, *Contours of evolution* are interesting for the history of orthogenetic concepts because they focuses on the search for ways of their development. The question of directionality was a key to the progress not only of biology but also humanities. Akhnazarov was aware of the fact that he could not create the concept of evolution of "everything" alone. He sought to develop an appropriate scientific program and to combine the efforts of experts from different sciences. Other authors of the concepts of directed evolution usually did the opposite, trying to counterpoise their work to that of everyone else.

Akhnazarov's interest in evolution was inspired by his observations on the escalation of management costs with the growth of production. With the increasing complexity of production, the complexity of the control system increases too, and even faster. The costs for its maintenance increase, and their impact on the efficiency of production is not negligible. Something similar seems to occur in the evolution of organisms. One of the main directions of animal evolution is cephalization, an increasing size and complexity of the nervous system and the brain. The brain of higher organisms consumes a huge amount of resources as compared to other body parts. What if some "costs" are involved here too? If so, where would evolution lead us further?

Akhnazarov emphasised that he presents not a "theory" but a "concept" or "approach". Its essence can be summarised as follows. The contradiction whose manifestation is the nature of evolutionary process is the contradiction between the basic properties of matter: "reflection" and "impact".[2]

At the inorganic stage of evolution these features are identical. At the biological stage these functions are separated within an individual, which is materialised in the formation of the respective body systems (the neural and receptor system and the muscular and effector one). The basic contradiction of matter becomes expressed as a contradiction between the ideal and the sensual activities ("the asymmetry of functions"). With time, the "reflection system", that is, the nervous system, becomes increasingly more isolated and complicated. In humans an asymmetry of the nervous system is present. Moreover, higher animals possess not only an "asymmetry in the space" but also a "temporal asymmetry", that is, an alternation of sleep and wakefulness. During sleep the nervous system seems to be engaged in something distant from the everyday life—it reflects, and after waking up it is engaged in everyday affairs—it impacts. In human society, these functions are separated at a social level as the division of labour.

Considering the directionality of evolution, Akhnazarov discussed the ideas of Schmalhausen, Gilyarov and several other authors and came to the conclusion that their mistake was to analyse directionality in separate evolutionary lineages or taxa. He believed that a more promising approach would be to focus on all living organisms and even on what preceded the origin of life.

Similarly to other authors supporting the idea of directed evolution, Akhnazarov analysed constraints on variation and parallelisms criticising "ectogenesis" and

---

[2] "... we identify these properties, analysing the most general characteristic of matter—its ability to interact, to enter into a relationship, to relate ... Impact is the ability of an object to pass its own properties to another object, in interaction with it, to "imprint" its own structure in its structure. Reflection is the ability of an object to reflect in its own structure, in interaction with another object, the basic properties (the system of connections and relations) of the latter. Reflection, facing the nature of the reflected object, and impact, facing the object as a given entity, a phenomenon, are related by the law of the unity of opposites, which is the primary contradiction of matter. The central evolutionary trend of the development of the contradiction between reflection and effect is a tendency for isolation of these properties from each other, their divergence, accompanied by a simultaneous enhancement of communication and interaction between them." (Akhnazarov 2002, p. 39).

"random evolution". Surprisingly, he found "autogenesis" unsatisfactory, too, feeling that there was some ambiguity in it. Like most theorists of evolution, he gave poor vitalism a kick: "All kinds of vitalism have been hopelessly compromised. The inner force of Teilhard de Chardin is, unfortunately, as non-constructive as Lamarck's 'pursuit of perfection'" (Akhnazarov 2002, p. 100). He looked for the way out of this situation in the idea of inheritance of acquired characters. "Therefore, abandoning the idea of random evolution and random mutations, we must make the next step: to relate mutagenesis with the impact of the environment, which means, eventually, to assume the inheritance of acquired characters in ontogenesis in one form or another. This way is clearly marked with a DANGER sign. But still it is a way forward." (Akhnazarov 2002, p. 100).

Akhnazarov tried to reinforce this bold statement by the hypothesis of L. B. Mekler postulating that the mechanism of the inheritance of acquired characters is associated with oncogenesis (Mekler 1980). This process undermines the internal structure of the body but does not disturb its morphological integrity, and therefore allows significant changes of the genome. At the same time, the Lamarckians of the past were mistaken in thinking that the direct influence of the environment caused changes in all organisms. In fact, only some of them undergo changes. If so, a limited reconciliation of Lamarckism and Darwinism might still be possible.

Akhnazarov was apparently aware of the weakness of this hypothesis. However, he thought it was important to find at least some explanation of Lamarckism from the viewpoint of modern biology. He still hoped that the mechanism of definite mutations caused by a direct impact of the environment would be found. These considerations may seem naive but it should be noted that Lamarckian ideas sometimes surface in modern immunology and molecular biology (see for review Golubovsky 2000).

References

Akhnazarov EB (2002) The contours of evolution (Kontury evolyutsii). Nedra, St Petersburg

Angelier E (1982) Albert Vandel (1894–1980). Bulletin de la Sociétė d'histoire naturelle de Toulouse 118: 21–23

Croizat L (1958) Panbiogeography. 3 vols. Published by the author, Caracas

Croizat L (1960) cited in Panke J, Danova N (2003) Leon Croizat, the founder of panbiogeography (Leon Croizat – osnovatel' panbiogeografii). In: In the shadow of Darwinism. Alternative theories of evolution in the 20th century (V teni darvinizma. Al'ternativnye teorii evolyutsii v XX veke). Yasnyy den', St Petersburg, pp 94–111

Croizat L (1962) Space, time, form: the biological synthesis. Published by the author, Caracas

Croizat L, Nelson G, Rosen DE (1974) Centres of origin and related concepts. Syst Zool 23:265–287

Edlinger K, Gutmann WF, Weingarten M (1991) Evolution without adaptation (Evolution ohne Anpassung). Aufsaetze und Reden Nr 37 der Senckenbergischen Naturforschenden Gesellschaft. Verlag Waldemar Kramer, Frankfurt am Main

Golubovsky MD (2000) The age of genetics: evolution of ideas and notions (Vek genetiki: evolyutsiya idey i ponyatiy). SPbFIIET RAN, St Petersburg

Grasshoff M (1994) The Frankfurt theory of evolution and the notions of “Anpassung” und “Selektion” (Die Frahkfurter Evolutions-Theorie und die Begriffe “Anpassung” und “Selektion”). Nat Mus 124(6):196–199

Grehan JR, Ainsworth R (1985) Orthogenesis and evolution. Syst Zool 34(2):174–192

Gudo M, Gutmann M (2003) Evolution: construction, reconstruction and mechanisms of evolution (Evolyutsiya: konstruktsiya, rekonstruktsiya i mekhanizmy evolyutsii). In: In the shadow of Darwinism. Alternative theories of evolution in the 20th century (V teni darvinizma. Al’ternativnye teorii evolyutsii v XX veke). Yasnyy den, St Petersburg, pp 176–193

Gutmann WF (1994) Constructional constraints in the evolution: swimming tetrapods (Konstruktionszwänge in der Evolution: schwimmende Vierfüsser). Nat Mus 124(6):165–189

Gutmann WF (1996) Are there alternative ways for the development of the organisation of living organisms I (Gibt es Alternativwege fuer die Entwicklung der Organisation von Lebewesen I). Nat Mus 126(8):250–262

Gutmann WF (1997) Globalization response in the area of classical biology and palaeontology (Globalisierung-Rückwirkung auf den Bereich der klassischen Biologie and Palaeontologie). Nat Mus 127(7):209–219

Gutmann M, Weingarten M (1992) Principles of construction morphology and organismic theory of evolution (Grundlagen von Konstruktionsmorphologie und organismischer Evolutionstheorie). Aufsaetze und Reden senck naturf Ges 38:51–68

Gutmann M, Weingarten M (1994) Changes in the discussion of evolutionary theory: the abolishment of atomism in genetics (Veränderungen in der evolutiontheoretischen Diskussion: Die Aufhebung des Atomismus in der Genetik). Nat Mus 124(6):189–196

Kaplan RW (1992) Is a new non-Darwinian theory of evolution necessary? The construction of organisms. I. Coherence, energy and simultaneous causality (Ist eine neue nicht-darwinistische Evolutionstheorie nötig? Die Konstruktion der Organismen. I. Kohärenz, Energie und simultane Kausalität). In: Gutmann WF (ed) Aufsätze und Reden Nr 38 der Senckenbergischen Naturforschenden Gesellshaft. Verlag Waldemar Kramer, Frankfurt am Main, pp 185–196

Kawamura K (2002) The origin of life from “the life of subjectivity”. In: Fundamentals of life. Éditions scientifiques et médicales Elsevier SAS, pp 563–575

Kawamura K (2003) The relativity of the gene, individuality and self-organisation in connection with the origin and evolution of life (Otnositel'nost' gena, individual'nosti i samoorganizatsii v svyazi s proiskhozhdeniem i evolyutsiey zhivogo). In: In the shadow of Darwinism. Alternative theories of evolution in the 20th century (V teni darvinizma. Al’ternativnye teorii evolyutsii v XX veke). Yasnyy den’, St Petersburg, pp 220–237

Lima de Faria A (1988) Evolution without Selection: Form and Function by Autoevolution. Kindle Edition

Lima de Faria A (1995) Biological periodicity: its molecular mechanism and evolutionary implication. JAI Press, Greenwich

Mekler LB (1980) A general theory of biological evolution. A new approach to the old problem (Obshchaya teoriya biologicheskoy evolyutsii. Novyy podkhod k staroy probleme). Zhurnal VKhO im DI Meneleeva 3:333–344

Panke J, Danova N (2003) Leon Croizat, the founder of panbiogeography (Leon Croizat – osnovatel’ panbiogeografii). In: In the shadow of Darwinism. Alternative theories of evolution in the 20th century (V teni darvinizma. Al’ternativnye teorii evolyutsii v XX veke). Yasnyy den’, St Petersburg, pp 94–111

Popov IY (2002) “Periodical systems” in biology (a historical issue). In: Hossfeld U, Th J (eds) The origin of biological disciplines (Die Entstehung biologischer Disciplinen). II. VWB, Berlin, pp 55–69

Reif W-E (1983) Evolutionary theory in German paleontology. In: Grene M (ed) Dimensions of Darwinism. Cambridge Univ Press, Cambridge, pp 173–203

Remane J (1983) Selection and the theory of evolution: should “old-Darwinian dogmas” be replaced by a critical theory of evolution? (Selektion und Evolutionstheorie: Müssen “altdarwinistische Dogmen" durch eine kritische Evolutionstheorie ersetzt werden?). Palaeontol Z 57(3–4):205–212

Schimkewitsch WI (1906) On the periodicity in the classification of pantopods (Über die Periodizät in dem System der Pantopoden). Zool Anz 30:1–22

Schindewolf OH (1936) Paläontologie, Entwicklungslehre und Genetik: Kritik und Synthese. Borntraeger, 1936

Schindewolf OH (1947) Questions of evolution (Fragen der Abstammungslehre). Verlag Dr. Waldemar Kramer, Frankfurt am Main

Schindewolf OH (1993) Basic questions in paleontology: geologic time, organic evolution and biological systematics. Univ of Chicago Press, Chicago

Teilhard de Chardin P (1950) Concerning a remarkable case of orthogenesis of a group: evolution of Chinese siphneides (Sur un cas remarquable d'orthogénèse du groupe: l'évolution des siphnéides de Chine). Paléontologie et transformisme. Albin Michel, Paris, pp 169–173

Teilhard de Chardin P (1987) The phenomenon of man (Fenomen cheloveka). Nauka, Moscow (Russian edition)

Vandel A (1965) Biospeleology the biology of cavernicolous animals. Pergamon Press, Oxford (English edition)

Vorobyeva EI (1991) Evolutionary synthesis and evolutionary morphology (Evolyutsionnyy sintez i evolyutsionnaya morfologiya). In: Vorobyeva EI, Vronsky AA (eds) Modern evolutionary morphology (Sovremennaya evolyutsionnaya morfologiya). Naukova Dumka, Kiev, pp 244–262

Chapter 7
Statements in Support of Orthogenesis in the 1950s and Later

Abstract In addition to new or revived ideas related with orthogenesis, there were in the recent history of evolutionary biology many statements in its support. Russian sources are especially interesting in this respect because in the Soviet Russia Darwinism occupied the same place in biology as communism did in society. Orthogenesis was in contrast with official ideology, and its support was fraught with serious problems. Yet orthogenetic heresy surfaced in the Soviet Union with a persistence that seems to indicate the solidity of its argumentation base. A similar phenomenon—the emergence of orthogenetic ideas under condition of dominating Darwinism—can also be seen in other countries. Though ideological pressure on biology was less dramatic there, non-Darwinian concepts were also considered as somewhat heretic—and yet they kept emerging.

Keywords Recent history · Support of orthogenesis · Heresics · Darwinism · Dissidents

In addition to new or revived ideas related with orthogenesis, there were in the recent history of evolutionary biology many statements in its support. Some were veiled but others were quite explicit. Russian sources are especially interesting in this respect because in the Soviet Russia Darwinism occupied the same place in biology as communism did in society. This situation established after 1929, the year of the "great turn" (*velikii perelom*) proclaimed by Stalin, which marked the total victory of socialism and the extermination of all capitalistic elements in the society. Censors looked avidly for any deviations from Darwinism in biological research, and accusations of non-Darwinism were likely to come up in any polemics on evolutionary problems. Orthogenesis was in too sharp a contrast with the official ideology to have a right for existence. In spite of this, orthogenetic heresy surfaced in the Soviet Union with a persistence that seems to indicate the solidity of its argumentation base.

Alexander A. Lyubishchev (1890–1972) was one of the most active dissidents in theoretical biology. He first expressed his ideas about evolution in 1925 (Lyubishchev 1925) and wrote about them for almost half a century after that, passionately criticising Darwinism and other biological concepts (Lyubishchev 1973a,

I. Popov, *Orthogenesis versus Darwinism*,
https://doi.org/10.1007/978-3-319-95144-7_7

b, c, 1975). While he added little new to this field, he did attract attention to its problems.

Estimations of Lyubishchev's scientific heritage are contradictory. Some authors admire him as a thinker and as a person, discuss the niceties of his ideas and praise the depth of his understanding of biological problems (see Svetlov 1982; Lyubishchev 1982; Chaikovsky 1984; Golubovsky 2000). Numerous conferences were organised and collections of papers were published in his honour. His ability to criticise everything was compared with the "coloured ear" of Vladimir Nabokov and other extraordinary human faculties (Golubovsky 2000). However, there is also an opinion (usually expressed in oral discussions) that Lyubishchev did not really suggest any new ideas and that his writings are too much like newspaper articles. A balanced viewpoint is that Lyubishchev's writings are related to the works of evolutionary theorists in the same way as literary criticism is related to literature, and should be considered correspondingly.[1]

Much of Lyubishchev's heritage is represented by letters and lectures rather than scientific publications. This was only partly due to the heretical character of his ideas. While Lyubishchev's writings were at odds with the official stance on theoretical biology, censorship became less stringent in the 1960s and some of Lyubishchev's papers were published in academic sources. They were, however, not as numerous as one might expect. His views on the natural selection theory were best expressed in the article *On the Postulates of Modern Selectogenesis* (Lyubishchev 1973a), where he supported nomogenesis, branded selectogenesis and stigmatised the inferiority of Darwinian thinking. Similarly to many other supporters of orthogenesis, Lyubishchev paid attention to similarities between inorganic structures and living organisms. One of his articles dealt with frost patterns on window panes. He speculated about their similarity with plants and suggested that the unity of laws governing growth and structure of objects was the only plausible explanation. The implication was that the same laws also determined evolution.

Though Lyubishchev disapproved of selectogenesis and supported nomogenesis, "severe criticism" of the latter can also be found in his writings (see Meyen and Shreider 1973, p. 40). Moreover, Lyubishchev claimed that he was actually a Darwinist. "Were Darwin to live up to our times and to keep a complete freshness of mind and the capacity for work, he would be in the camp of anti-Darwinians" (Lyubishchev 1973b, p. 45).

Out of the numerous words that Lyubishchev himself and his followers used to describe his position, "nihilism" seems the most appropriate. Such a stance is convenient but hardly constructive. Nevertheless, some of Lyubishchev's criticism cannot fail to appeal to readers interested in evolutionary biology. His wordings of the main errors and failures of speculations of evolutionists are quite remarkable. He named among them "use of unproved postulates as absolute truths; abuse of extrapolation; exaggerating the significance of experiments and ignoring descriptive facts; ascribing absolute priority to proof by contradiction and the principle of the excluded

[1] This viewpoint was expressed by Daniil A. Alexandrov at workshops of the Institute for the history of science of the Russian Academy of sciences in 1997.

middle" (Lyubishchev 1975, p. 206). Lyubishchev pointed out that in difficult cases the supporters of the modern evolutionary synthesis often appealed to philosophy or "found shelter in ignorance" (e.g. referring to the incompleteness of the fossil record). According to Lyubishchev, the architects of the modern synthesis, especially Ernst Mayr, thought that "a bad explanation was better than no explanation at all or an explanation inacceptable due to philosophical considerations" (Ibid., p. 210).

After Lyubishchev's death, his followers published numerous papers about him and his ideas. Conferences devoted to the development of Lyubishchev's heritage ("Readings from Lyubishchev") were organised. Proceedings of these conferences often contain statements in support of orthogenesis, which was usually understood as nomogenesis *sensu* Berg (see Svetlov 1982; Lyubishchev 1982; Chaikovsky 1984; Golubovsky 2000b).

Orthogenesis was supported in some publications aiming at reforming theoretical biology. Many of them employed the following line of argument: modern evolutionary synthesis faces insurmountable difficulties—its criticism from the position of orthogenesis is correct—yet orthogenesis, being somewhat unclear, is also inacceptable—a new theory should be created— for this, orthogenesis would be at least as useful, if not more so, as the modern synthesis. In particular, this logic can be seen in the works of **Yunir A. Urmantsev** (1931–2016), botanist and philosopher. Urmantsev criticised modern Darwinism and agreed with many of Berg's postulates. He claimed that his "general theory of systems" (Urmantsev 1974, 1979, 1988) was a modern alternative to these two concepts as it eliminated the need for various antitheses such as that between "polymorphism" and "isomorphism" (differences and similarities between organisms).

Sergei V. Meyen (1935–1987), a geologist and a palaeobotanist, actively developed the concept of directed evolution (mostly understood as nomogenesis) in the Soviet Russia. He put much effort into revision of the principles of taxonomy and evolutionary theory using the results of his own palaeobotanical studies. Meyen acknowledged that his views on evolution and taxonomy were influenced by Lyubishchev and Urmantsev, and his works resonate with both the general theory of system and Lyubishchev's ideas. Meyen simultaneously criticised and supported nomogenesis and other concepts of directed evolution (as well as the ideas of Lyubishchev and Urmantsev): "Nomogenesis as presented by L. S. Berg and D. N. Sobolev (1924) does not distinguish between static and dynamic laws, does not take into account the self-dependence of the laws of transformation of forms. This variant of nomogenesis is to a large extent based on facts pertaining to morphology rather than evolution. In the variant elaborated by Lyubishchev the dialectic relationship between of the nomo- and the tychogenetic component was identified but left unsolved" (Meyen 1974, p. 363). Meyen pointed out numerous "unclear and ineffective notions" such as "the internal law of development", "the great plan of development", "gradation principle", "the law of the aspiration to progress", "organic growth", "phylogenetic growth" etc. The "phylogenetic preformism" of Lyubishchev was also included in this list. However, Meyen noted that "although Lyubishchev had stressed that preformism must be considered simply as a constraint on possibilities and not necessarily as evolution in a definite direction, his ideas were later interpreted in the sense that he had rejected" (Meyen 1975, p. 75). Such speculations supposedly promoted evolutionary biology and, in particular, orthogenesis in the form nomogenesis.

Meyen tried to conduct a "dialectical analysis" of evolutionary biology. He often mentioned various antinomies: selectionism dominates in evolutionary biology but nomogenesis persists; directionality is determined by internal factors but external ones also act upon it; every ban implies certain requirements, while every requirement implies bans etc. As for the major antinomy, that of nomogenesis vs. tychogenesis, it is similar to the dialectic categories of "necessity" and "casualty": both these aspects of evolution exist and interact. Correspondingly, the difference between nomogenesis and Darwinism is not in the set of evolutionary factors they admit but in the leading role ascribed to one of them. Meyen considered the attitude to convergences as the main distinction between selectionism and nomogenesis.

In spite of this dialectical approach, Meyen apparently leaned towards nomogenesis, supporting it and other evolutionary heresies with numerous instances of parallelisms, aristogenesis, non-adaptive characters etc. At the same time, in every such case he emphasised the limitations of the concepts of directed evolution—the unclearness of the "laws". "The opponents of the selectionistic interpretation of directed evolution, while being justly indignant at its vagueness, follow the same way proposing such notions as "phylogenetic growth", "aspiration to perfection", "rectigradation" etc. They are far from being clear and cannot compete for the status of natural law having not only an explanatory force but a predictive one as well" (Meyen 1975, p. 67).

Meyen thought that "the general theory of systems" could solve all these difficulties, making "the laws" more concrete and reconciling Darwinism with its critics. "To eliminate the contradictions between nomogenesis and selectogenesis, it is necessary, first of all, to approach the variety and repeatability in the organic world from a position of the general theory of systems. This means the identification of systems at the intra-organismic, organismic and above-organismic levels. Primary units, the relations of unity and the laws of composition should be revealed within each of these systems. In such a way we will understand what must exist and what must not exist for these systems (i. e. a nomogenetic pattern of evolution) as well as what may or may not exist for them (this uncertainty represents a tychogenetic component of evolution). Secondly, distinguishing the nomogenetic and the tychogenetic component in every aspect of evolution, we should take into account their polymorphic (reciprocally multiple-valued) correspondence, i. e. to steer clear of reductionism of any kind. There are no reasons to consider contradictions between the two doctrines, selectionism and nomogenesis, as insurmountable." (Meyen 1974, p. 363).

In his last publications, Meyen expressed a different idea about the collisions in evolutionary biology and taxonomy: our knowledge is insufficient to construct any complete theory. He analysed several complex problems such as providing the rationale for the principles of phylogenetic reconstructions or the adaptive significance of characters. In every case it turned out that the existing methods of their solution were not well-grounded. His general conclusion was that we simply cannot be sure about anything in evolutionary biology. We know as good as nothing about the mechanisms of evolution. This should be openly admitted. It is no use trying to hide contradictions under enormous piles of empirical data. Evolutionary biology does not lack new facts, it lacks a theory. There are many facts but we do not know what to do with them (Meyen 1975, 1979, 1984, 1990).

Orthogenesis was also supported by another Russian scientist, **Leonid I. Korochkin** (1935–2006), who suggested the hypothesis of programmed transformation (Korochkin 1985). Korochkin cited Berg and Zavarzin and developed similar ideas based on the analysis of data on the molecular structure of genome. (To note, he stressed the preliminary character of his conclusions and the scarcity of data.) Korochkin analysed information about the repeated sequences of genes and found interesting parallelisms. For example, in respect of the content of repeated sequences in the genome, several genera of fishes are parallel to several genera of amphibians: the series *Salmo* – *Bufo* – *Necturus* is parallel to the series *Onchorhynchus* – *Xenopus* – *Triturus*. The patterns of the arrangement of unique and repeated sequences in the genome can be classified into three types: *clawed frog*, *fruit fly*, *albatross*. These types are scattered chaotically over the animal world. For instance, the *clawed frog* type was found in humans, loaches, sea urchins, primitive fungi and the house fly. The *fruit fly* type was found in the mosquito, the bee and the blow fly (which are relatively close to *Drosophila*), in pigeons and in some ground squirrels. The *albatross* type was found in the pelican, the guillemot, the Muscovy duck, some ground squirrels, turtles and pigs.

How can we look for the common ancestor of the house fly and the blow fly if their genome structure is so different? Such questions are difficult to answer within the modern evolutionary synthesis, and its advocates consider this direction of studies as a dead end. Korochkin cited Boris M. Mednikov: [this direction] "… works for the sake of itself… No taxonomist in his right mind would be interested in a character shared by an albatross, a turtle and a pig or by a chicken and a fruit fly or by a house fly and a clawed frog." (cited in Korochkin 1985, p. 139). Sane taxonomists may indeed be uninterested in these parallelisms but the fact of their existence remains. According to Korochkin's hypothesis of programmed transformation, it can be explained as follows. Initially, living forms are represented by a set of archetypes, and each archetype contains a genetic program ensuring the implementation of a definite direction of ontogenesis. Mutations provide material for the variation of ontogenesis within the limits of type-specific norm of reaction. They do not produce anything new but give a possibility to realise a genetic program which had been blocked by something or another (Korochkin 1985). In this way, the term "archetype", which had seemed hopelessly outdated, emerged anew in the history of biology. It agrees well with the concepts of directed evolution.

"Orthogenetic sympathies" can be seen in the concept presented by **Vitaly A. Kordyum**, which addressed the evolution of biosphere and highlighted the role of the horizontal transfer of genes in evolution (Kordyum 1982). Kordyum criticised Darwinism, employing many of the old arguments used by the supporters of orthogenesis, although he mainly used the material from modern disciplines such as molecular biology and the genetics of populations. Though Kordyum "was far from believing in orthogenesis", he still thought it was better than the modern evolutionary synthesis.

In some cases, the support of orthogenesis naturally followed from empirical observations. **Dmitri L. Grodnitsky**, studying the nevrulation of insect wings, found numerous convergences and/or parallelisms, which could hardly be explained by the natural selection theory (Grodnitsky 1995). The issue of the "internal or morphogenetic aspect" of evolution arose again. According to Grodnitsky, the recognition of the existence of this aspect ("structuralism") was to become an addition to Darwinism, perfecting the theoretical background for the analysis of evolution.

Another instance when empirical studies led the author to support the concepts of directed evolution was the morphological research of **Elena E. Kovalenko** (1953–2014), who suggested an alternative to Darwin's concept (2003). She wrote that "during the study of variation factual material was collected that did not correspond to Darwin's views on the character of variation" (Kovalenko 2003, p. 193–194). Darwin considered variation as random, unlimited and indefinite. In the process of her research, Kovalenko found that the number of actual variants was restricted, i. e. it was lesser than the theoretically possible number. It was illustrated with evidence from the sacrum of *Anura*: out of the 288 possible variants, 45 were found. At the same time, not only the number but also the set of variants turned out to be constant for the species. This set could be revealed in samples from different populations, in the offspring of various parents, in young and adult individuals. Moreover, the frequency of occurrence of certain variants was also constant. It seems that, given the lack of elimination and a high fecundity, the entire spectrum of the most probable variations might be registered in the offspring of two parents. This means that the idea of selection, i. e. selective elimination, is unnecessary for the explanation of natural variability. Selective elimination can be a particular case but is superfluous for the general evolutionary concept. "Under any conditions and under any kind of elimination the species will be represented by the variations that occur at a higher frequency in its individual variability." (Ibid., p. 207). It follows that only reproduction (copying of hereditary material) is required for the saturation of the gene pool of a species or its populations.

Kovalenko also used some data from dog breeding. She pointed out that the most variable breeds were those that were intensively bred. In these cases no selection pressure was created but all offspring (and, thus, all possible variations) simply were retained. At the same time, breeders are forced to change the standards of popular breeds continuously, suggesting numerous new subgroups. It would seem that breeds transform spontaneously, irrespectively of the breeders' efforts.

The main points of the "alternative theory" are summarised in Table 7.1. They have much in common with orthogenesis. Kovalenko claimed that previous concepts of directed evolution were a "brilliant criticism" rather than a real alternative to Darwinism. Though this statement is questionable, it should be admitted that Kovalenko indeed obtained interesting and unexpected data concerning variability as well as elaborated new approaches to the analysis of variation and new methods (see Kovalenko and Popov 1997; Zakharova 2002). These approaches and methods

Table 7.1 An alternative to Darwin's concept (Kovalenko 2003, p. 213)

	Natural selection theory	**Alternative to natural selection theory**
Cause of changeability	Changes in the environmental conditions; without them, there is no evolution	Impossibility to reproduce one's kind for an indefinitely long time; the fact of evolution does not depend on the environment
Role of variability (set of variations)	Passive source of changeability (the initial chaos of variations)	Determines the limits of changeability in a given time moment (sets the initial order)
Role of elimination	Active component of the mechanism of changeability: Actively creates the norm from chaotic variation; Directs the changes of the phenotype; Restricts the number of actual carriers of hereditary information.	Passive component: Restricts passively the range of actual variability compared to the potential one; Cannot influence the qualitative composition of "survivors" under conditions of the norm; Cannot decrease the range of potential variability.
Role of heredity	Passive (conservative) mechanism.	The most active mechanism creating new possibilities of the phenotype.
Intermediate forms (morphological series)	Result of a prolonged action of natural selection; reflect steps of successive transformations; must be extended in time.	Result of probabilistic pattern of development; do not reflect succession but follow the "either—or" principle; must emerge simultaneously.

take us closer to testing alternative viewpoints associated with orthogenesis vs. Darwinism polemics. The very suggestion that these viewpoints could be tested was a progress. Many authors of new concepts consider is sufficient to declare that the old ones are false.

Kovalenko emphasised that her alternative was born under the pressure of facts obtained from observations, without any theoretical influence. This must have been the case with other Russian supporters of orthogenesis as well because the Darwinism has traditionally been strong in Russia and the environment was not conductive to anti-Darwinian sentiments. At the same time, the emergence of orthogenetic ideas under condition of dominating Darwinism was not a characteristic feature of the Soviet/Russian science. The same phenomenon can be seen in the other countries where evolutionary biology developed. Though ideological pressure on biology was less dramatic there, non-Darwinian concepts were also considered as somewhat heretic. Yet, in spite of the enormous bulk of publications on evolution written in terms of natural selection theory, orthogenesis was from time to time recalled and found interesting or even worthy of use.

During this period, as before, orthogenetic ideas married especially well with palaeontology. A Dutch palaeontologist **Jan Hofker** came to support orthogenesis during his studies of extinct foraminifera. He showed series of transformations such as flattening of shell, increase of size etc. Forms described previously as separate genera turned out to be connected to each other with numerous intermediate forms. Hofker came to the conclusion that the borders of species and some genera were conventional. He believed that in most cases they were evolutionary stages of higher taxa. He proposed the term *gens* to denote the actual unity of the living world. A gens is a biological unit that starts at a certain moment from the species-level form, evolves directly through other forms and ends its existence as a final form, which is usually larger than the initial one (Hofker 1971). According to Hofker, the evolution of the *gens* was a display of a well-known phenomenon—orthogenesis. He believed that it could not be interpreted in terms of genetics because its cause, time, was hidden from experimentalists, and claimed that orthogenesis was simply an expression of time (Hofker 1959).

A similar interpretation of orthogenesis can be found in the study of **Hans Kaiser** dealing with anomalies (Kaiser 1970). He considered internally predetermined evolution as self-evident and hardly ever mentioned natural selection. Analysing "palaeopathology", Kaiser used the terms suggested by Schindewolf (typogenesis, typostasis and typolysis) though he did not share his concept completely. Kaiser was sceptical about Schindewolf's idea about the action of cosmic rays upon extinction. He believed that organisms reacted very differently to cosmic radiation and so it could not be a common factor behind extinctions. On the other hand, absolutely all "types" (*Bauplanen*) undergo endogenous "decentralization and disharmony of organs". Kaiser used the term "orthogenesis" in a sense different from that employed in this work. According to him, orthogenesis was the conservation of the trend of evolutionary changes, which occurred mainly for the benefit of organisms. Its opposite was senescence and death of taxa.

Some of the phenomena and problems highlighted by Kaiser are rarely addressed in modern biology. For instance, he argued, following Wilhelm Roux, that constancy was observed in the history of the Earth alongside with evolution. The study of its reasons would be enlightening. Unfortunately, noted Kaiser, it almost impossible to say something definite about the causes of constancy. The study of extinctions (typolysis) might contribute to the understanding of this phenomenon.

Kaiser took the idea of senescence and natural death of phylogenetic trends for granted. He analysed the ends of phylogenetic trends, that is, plant and animal groups with anomalous features as compared with the *Bauplan*. He used the term "anomaly" in a broad sense, applying it both to unhealthy conditions of organisms, to species which were, in his opinion, an indication of senescence or superspecialisation of trends and to some other phenomena. Kaiser specified that "anomaly" was synonymous with "deviation". He found "deviations" from the normal type in various extinct animals such as fanciful dinosaurs, Cainozoic mammals with enormous horns etc. They were considered as an evidence of "disharmony and decentralization" of the *Bauplan*. To describe the evolution of these *Bauplanen*, Kaiser used analogies with the organisms' ageing as well as with the onset of diseases.

Publications about orthogenesis were rather numerous in Spain. They were close to the ideas of Teilhard de Chardin. A Spanish author **Emilio Aguirre** wrote a voluminous article devoted solely to the analysis of orthogenesis. He said that while Mendeleev's system was periodic and rhythmic, that of Linnaeus was "aperiodic" and "arrhythmic". Most supporters of orthogenesis held an opposite view, trying to find periodicity and order in biology and stressing its similarity with physics and chemistry. On the contrary, Aguirre considered "aperiodicity" as an essential feature of the living world. He developed this idea illustrating it with various examples. For instance, he argued that even were it possible to construct biological system analogous to that of Mendeleev, it would contain numerous empty cells. Moreover, as there are many intermediate forms between groups of organisms, representing their diversity as a system of discrete units would be problematic. Biological variety was chaotic rather than periodic. No one could do in biology what Mendeleev did in chemistry, Aguirre argued.

Discussing evolution, Aguirre hardly ever mentioned the theory of natural selection, apparently considering it unworthy of notice. He thought that orthogenesis was the right way to knowledge about evolution. He pointed out the ambiguity of this term and tried to make it more precise. According to Aguirre, separating the two meanings, the descriptive and the explanatory one, was not enough. "Descriptive orthogenesis" had been used to describe at least two different phenomena: series of transformations in the evolution of organs or a reflection of biological progress. He analysed the relationship between progress and adaptation concluding that evolutionary trends (orthogenesis) were in general progressive, i. e. adaptive. He believed that progress and adaptation were the same phenomenon. Regress concerned only some aspects of evolution, while others, and so evolution in general, were progressive. However, he noted that adaptation could lead an evolving group to a dead end. An excessive development of some characters (*hypertelia*) caused extinction. The existence of evolutionary dead ends was in contradiction with the above statements about progress. Aguirre admitted this but argued that the deep mechanisms ensuring extension or narrowing of the organisms' capacities for adaptation were still poorly studied. He expressed a hope that the internal mechanisms of evolution would be explored in future (Aguirre 1958).

Another Spanish palaeontologist, **Miguel Crusafont-Pairó**, presented a different interpretation of directed evolution. He thought that the main evolutionary trend was Orthogenesis (from a capital letter), while all the other trends represented clinogenesis. In addition to the normal Orthogenesis, there was also a false or unsuccessful one (*ortogénesis frustrada*). The origin and evolution of birds was an example of this "frustrated orthogenesis". Birds have numerous progressive features such as homoiothermy, a four-chamber heart, feathers etc. and generally can be considered as a progressing group. Nevertheless, their evolution has been "erroneous" because, however many progressive features they may have, birds show no such progress of the nervous system as mammals and their ancestors. Crusafont-Pairó graphically described evolution from the lowest organisms to humans and revealed the one trend that resulted in the origin of man, i. e. Orthogenesis. According to Crusafont Pairó, "the ghost of finalism" must not frighten the researchers of this phenomenon (Crusafont Pairó 1960).

These works are reminiscent of old biological treatises. Orthogenesis has side branches but they are considered unimportant, and evolution in general is made to look like the *Scala Naturae* of Charles Bonnet (Bonnet 1745) or even Aristotle: the living nature is ordered from lower to higher forms, and Man resides on its top.

A more recent summary of orthogenesis was presented by **John R. Grehan** and **Ruth Ainsworth** (1985). These authors apparently discovered orthogenesis in the works of Leon Croizat and decided to enlighten the readers. Grehan and Ainsworth claimed that the contradictions of Darwinism and orthogenesis were overstated: the recognition of orthogenesis did not imply a complete rejection of natural selection. At the same time, the actual role of natural selection differed from the one postulated by Darwin. Organisms are predisposed to vary in certain directions, some of which are eliminated by selection while others are not. Grehan and Ainsworth noted that Darwin himself had not denied the "orthogenetic component of evolution" though they did not support this statement with sufficient evidence. They pointed out that the terms "bias", "constraints", "potential" in works describing evolutionary tendencies might imply orthogenesis.

A leaning towards orthogenesis can be seen in the theory of system prerequisites (*Systembedingungen*) of Rupert Riedl. The idea about the existence of order in nature lies at its core while the concept of natural selection is also considered correct. According to Riedl, Darwin's theory had to be supplemented by explanations of trans-specific evolution. Riedl pointed out several antitheses in evolutionary biology such as "necessity and casualty", "order and disorder" etc. and asked many questions such as "what is order?", "what is the problem of order?", "what are the principles of order?" etc. Clarifying his ideas, Riedl imagined numerous "disorders" in the structure of organisms and illustrated their impossibility with amusing drawings. He provided numerous formulas and often digressed into other sciences and philosophy. Unlike many modern authors, Riedl copiously cited his predecessors, who were also interested in the study of order in nature. He did not divide them into Darwinians and non-Darwinians. In his opinion, for instance, the works of von Baer, Waddington and Rosa contributed to the same trend of studies because they dealt with order in variation (Riedl 1975).

The publication of Schindewolf's book in English (Schindewolf 1993) provoked several new publications on orthogenesis. At least one of their authors, **Gerhard Rödder**, expressed support of typostrophism, although with some reservations (Rödder 1999). He believed that the modern evolutionary synthesis was based on such excellent studies that no criticism could ever harm it. If a new evolutionary concept was to be suggested in future, it would still remain within the existing theoretical framework. However, Rödder found numerous facts in his studies and in the modern literature that agreed well with the theory of typostrophism. For instance, he generalised recent data on the development of limbs characterising its molecular mechanisms and came to the conclusion that the stabilisation of a pentadactyl limb was caused not by any functional causes but by some internal ones. To explain these causes, Rödder used such terms as "inner-organismic coherence", "hardening" (or protection) "against modification by mutations". Rödder drew attention to the fact that the processes characterised by these terms corresponded to the typostasis of Schindewolf. He also recalled that Carson and Templeton in their studies of the fruit flies of Hawaii observed age-specific changes in populations: young ones were more capable of mutations. According to Rödder, this conclusion is basically identical to Schindewolf's statement about the senescence of phylogenetic trends (*Stammesalterung*).

Finally, Rödder presented his own data illustrating typogenesis, typostasis and typolysis. He composed diagrams of "origin, extinction, variety" of brachiopods and concluded that it was possible to agree, at least partly, with Schindewolf: "Tempo, extent, and diversity of modification are consequently grounded in internal factors of organic lineages and are not determined by the environment and its changes." (Rödder 1999, p. 199). At the same time, Rödder noted that little had changed in the global knowledge of the causes of evolution in the last 40 years.

In conclusion, it should be noted that small advances in the knowledge were often accompanied by large changes in the language. Coherences and hardenings replaced orthogenesis and typogenesis. Such coinage occurred even in reviews, whose authors should have been familiar with the original terminology. For example, **Dieter Korn** called the concepts of directed evolution "internalism" (Korn 2003). This term had been used before to denote an approach to the history of science within which only the scientific aspect is taking into account while social, ideological, psychological and other aspects are left out of the analysis. The use of internalism in the new sense would further confuse studies in the history of evolutionary theory.

A little later the concepts of directed evolution were enriched by another new term. Daniel McShea suggested an "internal-variance principle" to denote an increasing complexity of organisms in evolution (McShea 2005). He claimed that this phenomenon could not be explained by natural selection, though natural selection was, at the same time, the mechanism of adaptation. In spite of the fact that orthogeneticists usually criticised or altogether denied the adaptive nature of evolution, McShea wrote warmly about orthogenesis and thought that his internal-variance principle contributed to rehabilitation of this concept. He considered orthogenesis, the process driven by constraints on variation, as one of several evolutionary mechanisms. At the same time, his interpretation of orthogenesis was far from the primary sources. In his view, it was the heritage of Herbert Spencer and other philosophers. He did not refer any of the concepts of directed evolution mentioned above.

Some support of orthogenesis was also expressed outside biology. An approach of a non-specialist or a biologist trying to use the achievements of other sciences in biology often resulted in an idea that no chance events are possible in nature. Physics, chemistry and astronomy indicate order rather than disorder. If we consider evolutionary theory in this light, Darwinian view on adaptation originating from random variation would seem a nuisance. No chemist would say that copper originated from iron by selection of minute variations in the process of adaptation to a better melting temperature. Such explanations, however, are readily acceptable in biology. This absurdity led some theorists to look for the basis of evolutionary biology in Lamarckism and orthogenesis since they correspond better to the global laws found in other sciences. Non-equilibrium thermodynamics is especially popular in this respect because it indicates a regular and irreversible character of any historic process of transformation of matter. This approach is not discussed in the present work because the authors employing it seldom descend from the analysis of "everything" to the variation of living organisms and factors of evolution. "Macrodynamics does not study the mechanisms of phenomena. However in spite of these constraints, the macrodynamic method could be an effective tool of the study of ontogenesis, phylogenesis and biological evolution as a whole on a sound physical basis" (Gladyshev 1995, p. 13). Possibly, it would be possible to combine information on the directionality of the development of "everything" with the data about organisms in future. For the moment, however, we can just take into account that the evolution of organisms was a part of the global regular process, the evolution of the Earth and the Universe.

To sum up, statements in support of orthogenesis occurred throughout the period of the dominance of the modern evolutionary synthesis. They were often closely related with empirical studies. Although the concept of natural selection dominates in biology, some authors still try to reform it or even lapse into orthogenesis. The question of why that should be so is addressed in the following chapters.

References

Aguirre E (1958) Orthogenesis and the problem of biological evolution (La orthogenesis y el problema de la evolución biológica). Madrid Arbor 39 (148):467–502

Bonnet Ch (1745) A treatise on insectology or observations on some species of freshwater worms that, being cut into pieces, become whole animals again (Traité d'insectologie ou observations sur quelques espèces de vers d'eau douce, qui coupés par morceaux deviennent autant d'animaux complets). Durand, Paris

Chaikovsky YuV (1984) An analysis of the evolutionary concept (Analiz evolyutsionnoy kontseptsii). In: Systematicity and evolution (Sistemnost' i evolyutsiya). Nauka, Moscow, pp 32–53

Crusafont Pairó M (1960) The evolution of vertebrates. Orthogenesis and Clinogenesis (La evolución de los Vertebrados Ortogénesis y Clinogenesis). Estud Geol 16:1–23

Gladyshev GP (1995) On dynamic directionality of biological evolution (O dinamicheskoy napravlennosti biologicheskoy evolyutsii). Izvestiya AN Seriya biologicheskaya 1:5–14

Golubovsky MD (2000) The secret destiny of professor Lyubishchev (Taynyy zhrebiy professora Lyubishcheva). Vestnik 15(248):42–47

Grehan JR, Ainsworth R (1985) Orthogenesis and evolution. Syst Zool 34(2):174–192

Grodnitsky DL (1995) The problems of functional interpretation of some identical morphological structures on insect wings and an explanation of secondary similarity of organisms (Problemy funktsional'noy interpretatsii nekotorykh odinakovykh morfologicheskikh struktur na kryl'yakh nasekomykh i ob'yasnenie vtorichnogo skhodstva organizmov). Zh Obshch Biol 56(4):438–449

Hofker J (1959) Orthogeneses of foraminiferans (Orthogenesen von Foraminiferen). Neues Jb Geol Paläontol Abh 108(3):239–259

Hofker J (1971) An orthogenetic evolution of quinqueloculines from sigmolines to spirosigmolines (Une Evolution orthogenetique de quinqueloculina en passant par sigmolina vers spirosigmolina). Rev Micropaleontol 14(1):69–75

Kaiser HE (1970) The abnormal in the evolution (Das Abnorme in der Evolution). E. J. Brill, Leiden

Kordyum VA (1982) Evolution and biosphere (Evolyutsiya i biosfera). Naukova Dumka, Kiev

Korn D (2003) Typostrophism in Palaeozoic Ammonoids. Paläontol Z 77(2):445–470
Korochkin LI (1985) Parallelisms in the molecular organisation of the genome and the problems of evolution (Parallelizmy v molekulyarnoy organizatsii genoma i problemy evolyutsii). In: Molecular mechanisms of genetic processes (Molekulyarnye mekhanizmy geneticheskikh protsessov). Nauka, Moscow, pp 132–146
Kovalenko EE (2003) An alternative to Darwin's concept (Al'ternativa kontseptsii Darvina). In: In the shadow of Darwinism. Alternative theories of evolution in the 20th century (V teni darvinizma. Al'ternativnye teorii evolyutsii v XX veke). Yasnyy den', St Petersburg, pp 192–218
Kovalenko EE, Popov IYu (1997) A new approach to the analysys of properties of variability (Novyy podkhod k analizu svoystv izmenchivosti). Zh Obshch Biol 58(1):70–83
Lyubishchev AA (1925) The notion of evolution and the crisis of evolutionism (Ponyatie evolyutsii i krizis evolyutsionizma). Izvestiya Biologicheskogo NII pri Permskom gosudarstvennom Universitete 4(4):137–153
Lyubishchev AA (1973a) On the postulates of modern selectogenesis (O postulatakh sovremennogo selektogeneza). In: Vorontsov NN (ed) Problems of evolution (Problemy evolutsii), 3, Nauka, Novosibirsk, pp 31–57
Lyubishchev AA (1973b) Frost patterns on windows (Observations and reflections of a biologist) (Moroznye uzory na steklakh [Nablyudeniya i razmyshleniya biologa]). Znanie – sila 7:23–26
Lyubishchev AA (1973c) The notion of nomogenesis (Ponyatie nomogeneza). Priroda 10:42–44
Lyubishchev AA (1975) Towards classification of evolutionary theories (K klassifikatsii evolyutsionnykh teoriy). In: Vorontsov NN (ed) Problems of evolution (Problemy evolutsii), vol 4. Nauka, Novosibirsk, pp 206–221
Lyubishchev AA (1982) The problems of form, systematics and evolution of organisms (Problemy formy, sistematiki i evolyutsii organizmov). In: Meyen SV, Chaikovsky YuV (eds). A collection of papers. Nauka, Moscow
McShea DW (2005) The evolution of complexity without natural selection, a possible large-scale trend of the fourth kind. Paleobiology 31(2):146–156
Meyen SV (1974) On the proportion of the nomogenetic and the tychogenetic aspects of evolution (O sootnoshenii nomogeneticheskogo i tikhogeneticheskogo aspektov evolyutsii). Zh Obshch Biol 35(3):353–364
Meyen SV (1975) The problem of directionality of evolution (Problema napravlennosti evolyutsii). In: Itogi nauki i tekhniki. Seriya: Zoologiya pozvonochnykh. 7. Problemy teorii evolyutsii. VINITI, Moscow, pp 66–119
Meyen SV (1979) Can there be a winner in the discussion about nomogenesis? (Mozhet li byt' pobeditel' v diskussii o nomogeneze?). Priroda 9:114–116
Meyen SV (1984) The principles of historical reconstructions in biology (Printsipy istoricheskikh rekonstruktsiy v biologii). In: Systematicity and evolution (Sistemnost' i evolyutsiya). Nauka, Moscow, pp 7–32
Meyen SV (1990) Non-trivial biology (Netrivial'naya biologiya). Zh Obshch Biol 51(1)
Meyen SV, Shreider YA (1973) Biological paradoxes of A. A. Lyubishchev (Biologicheskie paradoksy A. A. Lyubishcheva). Priroda 10:38–42
Riedl R (1975) The order of living things (Die Ordnung des Lebendigen). Verlag Paul Parey, Hamburg and Berlin
Rödder G (1999) Is Schindewolf's "Grundfragen der Palaontologie" still worth reading? An apologetic evaluation of his work. Paläontol Z 73(1/2):195–201
Schindewolf OH (1993) Basic questions in paleontology: geologic time, organic evolution and biological systematics. Univ of Chicago Press, Chicago
Svetlov PG (ed) (1982) Alexander Alexandrovich Lyubishchev. Nauka, Leningrad
Urmantsev YuA (1974) The symmetry of nature and the nature of symmetry (Simmetriya prirody i priroda simmetrii). Mysl', Moscow
Urmantsev YuA (1979) Nomogenesis on similarity in the living nature (Nomogenez o skhodstve v zhivoy prirode). Priroda, 9:116–121

Urmantsev YuA (1988) Evolutionics or the general theory of development of systems of nature, society and thinking (Evolyutsionika ili obshchaya teoriya razvitiya sistem prirody, obshchestva i myshleniya). AN SSSR, Pushchino

Zakharova EYu (2002) An analysis of regularities of phenotypical variability of eyespots in the browns (Lepidoptera: Nymphalidae: Satyrinae) (Analiz zakonomernostey fenotipicheskoy izmenchivosti glazchatykh pyaten barkhatnits [Lepidoptera: Nymphalidae: Satyrinae]). Extended abstract of Cand. Sci. (Biol.) Dissertation. Ekaterinburg

Part II
Orthogenesis: Pro et Contra

Chapter 8
"Laws" in Biology

Abstract Various "laws", "regularities" and "rules" have been especially often found in palaeontology and biogeography. The adherents of the idea of directed evolution considered that these laws indicated the existence of order in the living nature similar to the order of chemical elements or the existence of internal forces driving evolution. On the contrary, authors opposing the concepts of directed evolution argued that there was no need to invent obscure internal driving forces since everything could be easily explained by the known evolutionary factors.

Keywords Laws · Rules · Inertia · Irreversibility · Specialization · Parallel series in variability · Hollow curve

The first thing that attracts attention in the arguments of supporters of orthogenesis is their claim to the existence of laws in evolutionary biology. Various "laws", "regularities" and "rules" have been especially often found in palaeontology. The adherents of the idea of directed evolution considered that these laws indicated the existence of order in the living nature similar to the order of chemical elements or the existence of internal forces driving evolution. On the contrary, authors opposing the concepts of directed evolution argued that there was no need to invent obscure internal driving forces since everything could be easily explained by the known evolutionary factors.

One of the first such laws was the "law of non-specialised ancestors" or Cope's rule (Cope 1896). It was originally presented as a sophisticated reasoning rather than a brief idea. According to this law, early stages of a group's evolution are represented by non-specialised organisms; as they evolve, specialisation intensifies and eventually narrows so much that the organisms lose the ability to adapt quickly to a changing environment and become extinct. Cope's rule has been often used in interpretations of historical changes in various groups. An increasing specialisation was considered either as a fatal predestination or as a manifestation of natural selection, favouring specialisation and leading, in time, to an evolutionary dead end.

Within a strictly Darwinian approach, a species can always adapt to a changing environment because at least some of its numerous variations should be suitable for new conditions. The fossil record, however, demonstrates the opposite: many large

I. Popov, *Orthogenesis versus Darwinism*,
https://doi.org/10.1007/978-3-319-95144-7_8

groups could not adapt to a new environment and became extinct. For example, something prevented the dinosaurs from adapting to the conditions of the Cainozoic era. A possible explanation of evolutionary dead ends is that the group existed under very specific conditions, and their sudden change did not leave enough time for the natural selection process. For example, some Mesozoic aquatic and flying reptiles inhabited upwelling zones, where water rich in microelements rose from the lower layers of the ocean to the surface. These zones were referred to as "seaside resorts" (Nesov 1991): they were extremely rich in fish because microelements increased biological productivity. (Peruvian and Chilean coasts are a modern example.) Relatively small-scale geological changes could rapidly eliminate or rearrange these habitats, in which case species associated with them would die. Other dinosaurs could also have perished because they were adapted to some exotic habitats.

However, this theory fails to explain the global disappearance of Mesozoic reptiles. At least some dinosaurs lived outside the "resorts" but became extinct nevertheless. Moreover, this explanation is tautological: the assumption of specialisation of extinct reptiles is confirmed by the fact that they became extinct, while the fact of their extinction is explained by specialisation. This vicious circle is difficult to notice, and the explanation seems satisfactory, especially as one can always say we do not know all the details of conditions under which extinct groups existed. The adherents of orthogenesis, however, were unsatisfied with this explanation. They suggested that evolutionary dead ends were due to the fact that organisms could not vary arbitrarily. They are predisposed to vary in certain directions, which do not always correspond to the requirements of the environment, or else, at some evolutionary stage they may cease to vary at all.

The "law of specialisation" is difficult to interpret. There are numerous cases when narrowly specialised groups gave birth to new ones, which represented new stages of evolution (Amadon 1943). For instance, fish-like ancestors of amphibians were probably very specialised organisms inhabiting shallow coastal waters (see Vorobyeva 1977). Should any researcher familiar with the law of the non-specialised ancestor visit the Earth hundred million years ago, he would have probably considered the ancestors of amphibians as a narrowly specialised group approaching extinction.

A definition of specialisation is rarely given in studies discussing this phenomenon. One of the few authors who did it was an American ornithologist **Dean Amadon** (1912–2003). According to him, (morphologically) specialised animals were those that had "one or more organs or structures greatly modified or developed as compared with other related species" (Amadon 1943, p. 134). This means that only rare instances of specialisation are related with a greater dependence on a specific environment such as the dependence of some insect pollinators on a certain plant species or the dependence of some parasite on the host. These cases suggest that specialisation in itself is a way towards an evolutionary dead end. However, Amadon thought that some highly specialised forms such as whales were "flexible" in respect of evolution. Therefore, it is unsurprising that new groups may originate from highly specialised forms. According to Amadon, the absence of missing links in the fossil record is explained by the fact that they were highly specialised and rare, and are now difficult to find.

This indicates that specialisation is a contradictory notion. Cope's rule cannot be used as an argument in the polemics concerning natural selection vs. orthogenesis. However, it points to a phenomenon which is difficult to explain—extinction. Its existence is an indirect evidence of certain limitations of variation.

A similar situation concerns Depéret's rule: large or gigantic animals always represent the final stages of a group's evolution, never its initial stages. It had long been considered that size increases with evolution. According to Darwinians, it is evident that natural selection favours large organisms. They are stronger and more successful in competition with their smaller relatives. According to the adherents of orthogenesis, this is far from being evident. Under certain conditions a large organism may have difficulties in finding enough food, while its smaller relatives could survive. Nevertheless, a tendency towards increasing size can be seen in several phylogenetic lineages.

Size may also decrease in evolution. While proboscidians generally tended to increase in size, dwarf elephants about 1 m tall evolved several times. Finally, most elephants—giant, dwarf and medium-sized alike—became extinct. Only two representatives of the erstwhile numerous group have survived to our times.

The authors of a large study on progressive evolution claimed that natural selection favoured a decrease in the size of organisms (Zotin and Zotin 1999). A decreasing size is a way to avoid an "unfortunate situation" in an ecosystem. "One may imagine that representatives of a species, plants or animals, gradually become smaller under the pressure of competition since small individuals have a chance to survive—be it because their main enemies loose interest in them, or because their food resources increase or because they can hide in case of danger" (Zotin and Zotin 1999, p. 226). Having performed complicated calculations characterising metabolism, the authors concluded that growing small provides an advantage. For instance, passerine birds are a flourishing group.

A decrease or an increase of the size of organisms in the course of evolution can also be explained by some other mechanisms. Both dwarf and giant forms are often confined to small islands (giant turtles, Komodo dragon monitors, anoa bulls etc.). It may seem that these islanders stand apart from the main trends evolutionary trends but several evolutionists have attached great importance to them. The theory of isolation, dating back to the nineteenth century, considers dwellers of islands or isolated valleys as an indication of the only evolutionary way. Ernst Mayr, one of the authors of modern evolutionary synthesis, was especially active in promulgating this concept. According to him, were it not for islands or isolated areas within continents, no significant evolutionary events would have taken place. Mayr explained the differences of insular forms from their mainland relatives by the founder's effect: insular populations are founded by a few individuals carrying a small part of the species' gene pool. He believed that in the offspring of such individuals mutations rapidly arrive at a homozygous condition, become exposed to natural selection and can bring about genetic revolution (Mayr 1954). Usually, however, the origin of insular forms is explained by the genetic drift, which results in the emergence of rare variations, or by specific ecological conditions such as the scarcity of resources, the absence of predators or competition etc., which cause particular selection pressures.

All these speculations seem convincing enough. Yet in some cases no explanation for an increasing or a decreasing size can be found. In a study of sizes of iguanas—large and small, insular and mainland—no correlations between differences in size and differences in environment were found (Petren and Case 1997). Iguanas living on islands could be large or small under similar conditions such as the absence of competitors and predators, the variety of resources etc. Nevertheless, the authors mentioned the existence of certain evolutionary tendencies and constraints (ecological, physiological and genetic). In general, it is unclear whether an increase or a decrease in size is the main trend of iguanas' evolution since they are highly variable in this respect (Petren and Case 1997).

So, the size of organisms may increase or decrease in evolution. Both these processes can be considered as an adaptation, as a realisation of evolutionary tendencies or as a result of other mechanisms. Under close consideration, it seems that some groups "chose" the way towards increasing size while others "chose" the opposite way. This is hardly a satisfactory explanation. Certainly the organisms cannot choose an evolutionary pathway and start to evolve in the chosen direction? However, this is precisely the impression one gets from studying their morphology.

The "law of relative growth" or "allometry" is sometimes mentioned in connection with Cope's rule and Depéret's rule. These terms were coined be Julian Huxley to characterise the relationships of sizes of different parts of an organism during its growth. Evolutionary dead ends can be explained by allometry as follows: an increase in the size of an organ under the pressure of selection results, if allometric relationships are conserved, in an inadaptive growth of other organs. This explanation, however, did not seem convincing enough even for the supporters of the natural selection theory because natural selection acts not only upon separate organs but on the entire organism, i. e. the entire system of allometric relations (Jepsen 1949; Krasilov 1977). Critics of Darwinism gloated over this circumstance. In particular, P.-P. Grassé (1977) pointed out this fact as he criticised J. Huxley. However, Grasse found it difficult to interpret palaeontological rules too. In his opinion, nothing certain was known about their mechanisms.

Several authors pointed out another law of palaeontology—that of inertia. According to this law, once an evolutionary trend has manifested itself, it progresses until the end even if it leads in a non-adaptive direction. In other words, after reaching an adapted condition organisms continue to evolve by inertia. In a case of increasing size the law of inertia means that the size would continue to increase even after reaching the optimum. An Austrian palaeontologist **Othenio Abel** (1875–1946), who suggested the law of inertia (*Trägheitgesetz*), mostly relied on evidence from the evolution of odd-toed ungulates (Abel 1928, 1929). A certain evolutionary direction manifested itself early in the evolution of horse ancestors: the reduction of the number of fingers and an increase of size. This direction progressed irrespectively of anything and reached a culmination when an only finger was left and the size increased significantly. A similar inertia can be seen in the changes in the proportions of limbs, the structure of molars and other organs. All these changes occurred under various conditions, in forests as well as on grasslands. To note, Darwinists considered the evolution of horses as an adaptation to running and to the life in grasslands. They believed that climate changes resulted in the origin of grasslands such as steppes and savannahs and the changes were caused by new selection pressures arising under the new circumstances (Kowalewsky 1875).

In this case the final evolutionary stage seems to be either perfect or at least neutral. However, trends may end in bizarre forms, which seem to possess unnecessary if not downright harmful characters. Such cases, which have been referred to as hypertelia (Grassé 1943) or grotesque structures (Rensch 1954), are a better illustration of the law of inertia. A much-cited example of this kind is the extinct Irish elk (*Megaloceros giganteus*). It looked like the modern deer but was much larger and had enormous antlers, with the distance between their ends sometimes exceeding three meters (Fig. 8.1). For some reason, scientists tend to think that these antlers were unnecessary and eventually caused the downfall of their owners. They were too large, required much calcium and were, in general, a nuisance. Nevertheless, the elks apparently could not loose these superfluous antlers in the process of evolution.

We would never know whether the antlers were useful for the Irish elk (there are none left to ask) but it is probable that they were, at least, not harmful. Some modern

Fig. 8.1 Skull of the Irish elk, Musée d'Archéologie nationale, Saint-Germain-en-Laye

deer also have large antlers but they only affect their survival because human hunters like such trophies. Irish elks had existed for a long time over vast areas: their remains were found in Europe, Northern Africa and Asia as far as China. They became extinct quite recently, several thousand years ago. There are no reasons to assume that the antlers were the cause. Numerous other mammals species became extinct, including ungulates with "usual" antlers or without any antlers altogether.

However, if the Irish elk indeed disappeared because of antlers, it would be an excellent case for the law of inertia. The evolutionary trend was as follows: the antlers continued to increase in size and the deer became disproportionate, non-viable animals. "The misnamed, mistreated, and misunderstood Irish elk", wrote Steven Gould (Gould 1977b). It was called the Irish elk (= Irish moose) but it was a deer rather than a moose and it inhabited not only Ireland. Moreover, this elk-moose-deer suffered numerous erroneous interpretations of its evolution. Gould rejected them resolutely replacing them with the "correct" one: the Irish elk became extinct because it failed to adapt to the Subarctic conditions of the last glacial period (Gould 1977b). This explanation, however, is also hardly convincing. Irish elks inhabited a vast territory. They did not have to die out; instead, they could have reduce in numbers, or migrate southwards, or evolve into another species. Actually, it cannot be ruled out that Irish elks were exterminated by humans (Steinmann 1908).

Another palaeontological law introduced by Abel was Dollo's law or the law of the irreversibility of evolution. Abel referred to the following statement by Dollo: "…an organism cannot return, even partially, to a previous state already realized in its ancestral series." (Dollo 1893, p. 165). Numerous aquatic reptiles and mammals did not revert to the state of their fish-like ancestors but acquired new features suited for life in water. Dollo enthusiastically described numerous instances of such secondary adaptations such as quadrupedalism in some dinosaurs, arboreal lifestyle in some kangaroos, aquatic lifestyle of turtles etc. All these cases, however, can also be interpreted in a different manner. If we pay attention not to the differences but to the similarities of the secondary and the ancestral features, evolution might seem to be reversible: whales and ichthyosauri are fish-like, the flippers of turtles are similar to fins etc. Most supporters of directed evolution interpreted Dollo's law in favour of their viewpoint: evolution follows its own internal laws. Whales and ichthyosauri did not suddenly become suicidal and decide to drown themselves. They acquired characters that made their life on land unbearable such as short extremities and an elongated spinal column following the laws of variation. In this way, they were forced back to the environment of their remote ancestors.

Darwinists have often mentioned Dollo's law too. Apparently, they found it easy to interpret. From their viewpoint, organisms tend to adapt to any environment including that similar to the environment of their ancestors. Since both the environment and organisms continuously change in various directions, an exact repetition of previous evolutionary stages is statistically impossible.

Dollo himself reworded his law several times, while his followers suggested further improvements. In a recent work, it has been proposed to replace Dollo's law with "a broader and more exact law of the non-identity of similar structural transformations in different taxa of living organisms" and to replace the principle of irreversibility by that of contra-directionality of evolution (Seravin 1999, p. 89). No exact definitions were given however. In essence, apparently, it was implied that some evolutionary changes were reversible while others were not.

All biological laws are pertinent to discussions concerning orthogenesis. Dollo's law, however, cannot be considered as an argument because it is even more obscure than the previous ones. It should be noted that several authors interpreted the law of irreversibility of evolution in a different manner applying it only to the loss of organs. The most concise wording of this interpretation was given by Pierre-Paul Grasse: regressive orthogenesis is irreversible (Grassé 1943). In the evolution of elephants relatively small animals evolved into giants, which in turn gave rise to some small forms. No new organs appeared or disappeared, and evolution was reversible. However, once the number of fingers was reduced in the evolution of ungulates, it could not become larger at its final stages. There is not a single instance of this kind in the fossil record though the evolution of ungulates is well known. Their ancestors populated vast areas for a long time, and there are good reasons to believe that conditions favouring a greater number of fingers were formed somewhere at some moment. Yet the number of fingers only decreased and never increased. In terms of the natural selection theory, the number of fingers may increase under special selection pressure. However, experiments on breeding multi-fingered sheep, pigs, horses etc. are unlikely to be successful. The law of irreversibility seems to be correct in this sense and can be considered as an argument in favour of orthogenesis.

Besides palaeontology, laws have been found in other areas of biology. Sometimes they were closely connected with paleontological laws. For example, inertia was described in some extant populations of the moufflon (*Ovis gmeleni*). Even when predators were completely absent, moufflons continued—by inertia—to hide and gathered in groups, ready to defend themselves. Some salamanders (*Ambystoma barbouri*) demonstrate ineffective behavioural responses to predators, which are regarded as an ancestral feature persisting by inertia. Delayed maturation of some New World birds may also be due to inertia (Blomberg and Garland 2002). To note, in all these cases inertia did not result in extinction.

Another source of biological laws were data from taxonomy and biogeography. The works of Nikolai I. Vavilov are especially demonstrative in this respect since he made generalizations in both areas, describing series in hereditary variation and the centres of origin of cultivated plants. The law of homologous series in hereditary variability was used to support orthogenesis especially often. It refers to "regularities gleaned from the study of forms in the plant world" (quoted after Vavilov 1968, p. 13). Vavilov summarised these regularities as a formula (Fig. 8.2) and a set of tables (Fig. 8.3). Letters "L" indicated "radicals", the properties of Linnaean species; each is represented by similar versions of variability. This can be seen as an indication of the existence of uniform laws of variability and thus evolution. In some cases, Vavilov predicted the discovery of previously unknown variants of cultivated plants just as Mendeleev predicted the discovery of new elements. However, this law has been hardly ever used as a basis for taxonomy, and later studies in this direction added little new to it.

$$L_1(a+b+c+d+e+f+g+h+i+k+\ldots),$$
$$L_2(a+b+c+d+e+f+g+h+i+k+\ldots),$$
$$L_3(a+b+c+d+e+f+g+h+i+k+\ldots).$$

Fig. 8.2 Formula of Vavilov's law. L – "radicals" of Linnaean species; a, b, c … – races within species

One of the rare works analysing the possibility of using Vavilov's law as a biological "periodic table" was the paper by **Mikhail G. Popov (1893–1955)**, a botanist and an explorer of Central Asia. In his monograph about the genus *Cicer* (Leguminosae) he suggested that taxonomy required a "theoretical system" so that a researcher could follow certain known theoretical principles instead of incessantly revising the work of his predecessors. Vavilov's law seemed to be a suitable tool for this. In this respect, Popov considered two questions: 1) whether within the racial variation of a species forms absent in a closely related species could appear; 2) to what extent the law is applicable to taxa above the species rank. The first question remained largely unanswered. It was suggested that such forms do occur, but "it is completely unclear what their character is and how often they originate" because of the lack of material (Popov 1929). The second question was answered in a more definite manner. The closest genus to *Cicer* is *Vicia*. It is much more diverse than *Cicer*, and Popov found it inconceivable that numerous species similar to those in the genus *Vicia* would ever be discovered in *Cicer*. At the same time, among the relatively few forms in the genus *Cicer* there those unknown in *Vicia*. Popov concluded that it was impossible to use the law of homologous series for the construction of a theoretical system for the genera of Leguminosae.

"Laws" revealed in the geographical distribution of organisms by Vavilov and others are also related, though indirectly, with the polemics concerning orthogenesis. One of the major generalisations in this field is the "age and area hypothesis" or "space and time" concept by **John Christopher Willis** (1968–1958). Its essence is as follows: the size of the distribution area of a species is directly proportional to its age; widespread species evolved a long time ago, while species with a narrow distribution evolved recently and have not had time to spread far from the centre of their origin. According to Willis, the dispersal of a species in nature is slow since its numerous representatives need to overcome geographical barriers and colonise territories already occupied by other species. Evolution is seen as a "mechanical" process of spontaneous origin and dispersal of species, which has little to do with natural selection.

Vavilov's data on the centres of origin were consistent with the concept of Willis (Vavilov himself noted this fact). Vavilov put much effort into travelling to the centres of origin of cultivated plants (Vavilov 1968). In this way, he could get material for selection because the maximum diversity of a species is concentrated in the centre of its origin; as one moves to the periphery of the distribution area, the variety (that is, the ability to generate new races, strains, morphs etc.) decreases. For example, rye is cultivated primarily in Russia and northern Europe. It would seem that its

A general scheme of variability of cultivars (races) in the fam. Gramineae

Characters with hereditary variability	*Secale cereale*	*Triticum sativum*	*Hordeum sativum*	*Avena fatua* sens. ampl.	*Panicum mileaceum*	*Andropogon sorhum*	*Zea mays*[a]	*Oryza sativa*	*Argopyrum repens*
Inflorescence									
Articulation of spikelets and flowers (tendency towards self-sowing):									
spikelets and flowers sloughing during ripening (rachis brittle) — type *spontaneum*............................	+	+	+	+	+	+	+[b]	+	+
spikelets and flowers not sloughing (rachis non-brittle)[c].........................	+	+	+	+	+	+	+	+	+
Aristulate condition:									
grain aristulate (firmly enclosed in glumas or lemmas); not sloughing during ripening................................	+	+	+	+	+	+	+	+	+
grain naked (easily shedding glumas or lemmas during thrashing); easily sloughing during ripening.................	+	+	+	+	+	+	+	+	—
Density:									
dense..	+	+	+	+	+	+	+	+	+
loose...	+	+	+	+	+	+	+	+	+
intermediate......................................	+	+	+	+	+	+	+	+	+
Sex characters:									
unisexual plants................................	+[d]	—	—	—	—	—	+	—	—
bisexual plants..................................	+	+	+	+	+	+	+	+	+
Spinosity:									
spikelets awned (awns on lemmas)....	+	+	+	+	—	+	—	+	+
spikelets awnless...............................	+	+	+	+	+	+	+	+	+
spikelets short-awned and half-awned..	+	+	+	+	+	+	—	+	+
spikelets with deformed awns (type *furcatum*).....................	+	+	+	—	—	—	—	—	—
spikelets with awnlike appendages on glumas...	+	+	+	—	—	—	—	—	—
Characters of awns:									
a) rough...	|	|	+	+	—	+	—	+	+
soft..	+	+	+	+	—	+	—	+	+
b) indented....................................	+	+	+	+	—	+	—	+	+
smooth...	—	+	+	+	—	—	—	+	—
Number of flowers in spikelet:									
single-flowered spikelet....................	+	+	+	+	+	+	+	+	—
two-flowered spikelet........................	+	+	(+)	+	+	+	+	+	+
many-flowered spikelet......................	+	+	+	+	—	—	—	—	+

[a] Variations not only of ears but also of male inflorescences are taken into account.

[b] In teosinte (*Euchlaena mexicana*) interbreeding with Zea mays and in hybrids between maize and teosinte.

[c] Besides, in rye, barley, wheat and a number of species of *Aegilops* the falling off of spikelets only from the upper part of the internode (type *Aegilops cylindrica*) was registered.

[d] Found by V.F. Antropova in case of inzucht of cultivated rye in 1930.

Fig. 8.3 Table of variability rows of cereals as an illustration of Vavilov's law

diversity in these regions should be high—it is actively cultivated and selection work should be in progress. In practice, however, its diversity there is low. On the contrary, in Central Asia, where rye is little more than a weed of wheat fields, its diversity is the highest. Rye must have originated somewhere in Central Asia; afterwards its strains spread in various directions losing diversity in the process.

Similar ideas about the loss of diversity in the process of dispersal were expressed by other researchers: Joel Allen, who studied geographic variation of mammals and birds (Allen 1906), and William Reinig, who studied insects (Reinig 1938). The loss of diversity explained some zoogeographical rules but it remained unclear why it was lost in the first place.

In addition to the age and area hypothesis, Willis discovered the phenomenon of the "hollow curve" (Willis 1922, 1940). When flora and fauna of an area are characterised, taxa tend to form groups of unequal size: most genera have a single species, while the number of genera rich in species is minimal. A similar pattern is observed when one compares the distribution areas of species (Willis 1922, 1940). Willis mostly studied distribution and biogeography but the "hollow curve" also describes the grouping of organisms into taxa. The more thoroughly a group is studied, the more asymmetric its classification becomes: some genera within a family become larger and are divided into sections, subgenera and superspecies, while other genera remain small. The same phenomenon is observed within genera: numerous subspecies, morphs, stocks etc. are described within some species, while others remain relatively homogeneous.

The "hollow curve" seems counterintuitive. The genus as a taxonomic category should integrate related species. What use is a genus with a single species in it? It is much more convenient when different genera comprise an approximately equal number of species, and that is what taxonomists usually try to achieve. Correspondingly, the ideas of Willis excited little enthusiasm in his colleagues. They were criticised, sometimes quite offensively, in such terms as "nonsense", "speculations", "patterns that can be found in the telephone directory" etc. The latter accusation confused Willis greatly. It was precisely his aim to find a universal law operating outside botany and biology. He found similar patterns analysing the distribution of family names in the Swiss Canton of Vaud. A negative stance of taxonomists on these investigations was evidently due to the fear that the hollow curve would inconvenience their work. Indeed, the taxonomy of any group looks like an attempt to hush up the existence of the hollow curve. Yet however much would the taxonomists like to put the organisms into the Procrustean bed of a traditional classification, the hollow curve continues to resurface in modern studies. It can even be observed for classes and phyla, most of which contain just a few species or a single one.

A good example of the "hollow curve" is a classification of the fish family Coregonidae that was in use until recently. Many researchers agreed that there were about 27 species and 3 genera in the family. One genus (*Stenodus*) included a single species (inconnu or white salmon), the genus *Prosopium* included 6 species, while the rest of the species belonged to the genus *Coregonus* (whitefish). The latter has an extremely complex structure. It contains groups of species of unequal size, which can be described as "subgenera", "superspecies" or "bunches of species". There are several "good" species within *Coregonus* but several others are notoriously difficult to deal with. For instance, dozens of distinct forms, ecological as well as geographical, can be described within the species *C. lavaretus* (whitefish), which is distributed over a vast area of the Eurasian north. Sometimes these forms were described, not unreasonably, as separate species but since splitting was unpopular in the past, they were usually included into one species (see Reshetnikov 1980; Svärdson 1998). Should the rank of intraspecific whitefish forms be raised to the species, it would necessitate raising the rank of *Coregonus* species to genus, then raising the rank of genera to families (i. e. dividing the family of whitefishes into three families), then reviewing the entire system of Salmoniformes. Thus, the family Coregonidae seems to be in a permanent need of a complete revision (Popov and Sendek 2003).

Under close scrutiny, the hollow curve illustrates an entirely non-Darwinian scheme of evolution. It is not a tree whose branches grow in all directions but rather a horsetail: the main stem bears bunches of branches, the lowest of which had time to grow longer while the top ones have just started to grow. If we supplement this picture with information about parallelisms and homologous series, the entire scheme of evolution and taxonomy would be increasingly more suggestive of the table of chemical elements, that is, the existence of order in the living world.

References

Abel O (1928) The biological law of inertia (Die biologishe Trägheitsgesetz). Biologia generalis 4:1–102

Abel O (1929) Palaeobiology and phylogenesis (Palaeobiologie and Stammesgeschichte). Gustav Fischer, Jena

Allen J (1906) The influence of physical conditions in the genesis of species. Annual report of Smithsonian Institution (1905), Washington, 375–403

Amadon D (1943) Specialisation and evolution. Am Nat 77:133–141

Blomberg SP, Garland T (2002) Tempo and mode in evolution: phylogenetic inertia, adaptation and comparative methods. J Evol Biol 15(6):899–910

Cope ED (1896) Primary factors of organic evolution. Open Court Publishing, Chicago

Dollo L (1893) The laws of evolution (Les lois de l'évolution). Bulletin de la société belge de géologie, de paléontologie et hydrologie 7:164–166

Gould SJ (1977) The misnamed, mistreated, and misunderstood Irish elk. In: Norton WW (ed) Ever since Darwin, New York, pp 79–90

Grassé P-P (1943) Evolution. Facts, experiments, theories (L'Evolution. Faites, expériences, theories). Tournier & Constans, Paris

Grassé P-P (1977) Evolution of living organisms. Evidence for a new theory of transformation. Academic Press, New York

Jepsen GL (1949) Selection, "Orthogenesis", and fossil record. Proc Am Philos Soc 93(6):479–501

Kowalewsky WO (1875) The osteology of two extinct species of ungulates (Osteologiya dvukh iskopaemykh vidov iz gruppy kopytnykh). Izvestiya obshchestva lyubiteley estestvoznaniya, antropologii i etnografii, 16:1–59

Krasilov VA (1977) Evolution and biostratigraphy (Evolyutsiya i biostratigrafiya). Nauka, Moscow

Mayr E (1954) Change of genetic environment and evolution. In: Huxley J, Hardy AC, Ford EB (eds) Evolution as a process. Allen and Unwin, London, pp 157–180
Nesov LA (1991) Pterosauri over plane-tree forests and brackish-water bays of seas (Krylatye yashchery nad platanovymi lesami i solonovatymi zalivami morey). Gerpetologicheskie issledovaniya, 1:147–163
Petren K, Case T (1997) A phylogenetic analysis of body size evolution and biogeography in chuckwallas (Sauromalus) and other Iguanines. Evolution 51(1):206–219
Popov MG (1929) The genus Cicer and its species (Rod Cicer i ego vidy). Trudy po prikladnoy botanike, genetike i selektsii 21(1):1–240
Popov IYu, Sendek DS (2003) The quintessence of evolution (Kvintessentsiya evolyutsii). In: Evolutionary biology: history and theory, 2 (Evolyutsionnaya biologiya: istoriya i teoriya, 2). SPbFIIET RAN, St Petersburg, pp 172–189
Reinig WF (1938) Elimination and selection (Elimination und Selektion). Verlag von Gustav Fischer, Jena
Rensch B (1954) New problems of evolution: transspecies evolution (Neure Probleme der Abstammungslehre: Transspeciphische Evolution). Enke, Stuttgart
Reshetnikov YuS (1980) Ecology and systematics of coregonid fishes (Ekologiya i sistematika sigovykh ryb). Nauka, Moscow
Seravin LN (1999) The principle of contra-directionality of morphological evolution and the analysis of Dollo's law (Printsip protivonapravlennosti morfologicheskoy evolyutsii i analiz zakona Dollo). In: Evolutionary biology: history and theory (Evolyutsionnaya biologiya: istoriya i teoriya). IIET RAN, St Petersburg, pp 81–93
Steinmann G (1908) Geological foundations of the evolutionary theory (Die geologischen Grundlagen der Abstammungslehre). Engelmann, Leipzig
Svärdson G (1998) Postglacial dispersal and reticulate evolution of nordic coregonids. Nord J Freshw Res 74:3–32
Vavilov NI (1968) The law of homological sereis in hereditary variability (Zakon gomologicheskikh ryadov v nasledstvennoy izmenchivosti). In: The classics of Soviet genetics (Klassiki sovetskoy genetiki). Nauka, Leningrad, pp 9–51
Vorobyeva EI (1977) Morphology and characteristics of evolution of crossopterygian fishes (Morfologiya i osobennosti evolyutsii kisteperykh ryb). Trudy PIN, 163. Nauka, Moscow
Willis JC (1922) Age and area. Cambridge Univ Press, Cambridge
Willis JC (1940) The course of evolution by differentiation or divergent mutation rather than by selection. Cambridge Univ Press, Cambridge
Zotin AI, Zotin AA (1999) Direction, rate and mechanisms of progressive evolution (Napravlenie, skorost' i mekhanizmy progressivnoy evolyutsii). Nauka, Moscow

Chapter 9
Species Senescence

Abstract Many orthogeneticists argued that species and other taxa undergo the same transformation as individuals: they are born, develop and die. While the environment or competitors can indeed cause extinction, senile and weak groups, which are about to die anyway, are the first to succumb. Their opponents claimed that extinction was always a forced process caused by competitors and/or changing conditions and/or direct extermination and that the origin and evolution of species was not age-related. A lack of convincing explanations of past extinctions testifies in favour of species senescence.

Keywords Extinction · Senescence · Survival · Abiotic and biotic factors · Bioevents

Many orthogeneticists argued that species and other taxa undergo the same transformation as individuals: they are born, develop and die. While the environment or competitors can indeed cause extinction, senile and weak groups, which are about to die anyway, are the first to succumb. Their opponents claimed that extinction is always a forced process caused by competitors and/or changing conditions and/or direct extermination and that the origin and evolution of species is not age-related. Within the strict Darwinian approach, the historical development of a group is primarily explained by competition: the group flourishes if there are no stronger competitors and degrades and dies when they appear (Davitashvili 1948, 1969).

However, this explanation is often insufficient. The fossil record demonstrates that new groups do not force their predecessors out but simply occupy space made free after their extinction. So, Sobolev noted that broadly distributed groups often become extinct without the involvement of any competitors. For instance, the evolution of nautiloids was unrelated with that of fishes. Nautiloids coexisted with fishes and other aquatic animals for millions of years. Then they suddenly died out while ammonoids, which were quite similar, survived. Modern Darwinists explain this phenomenon as follows: we cannot understand what characteristic of the environment became the crucial aspect of the flourish or the decline of a group because we cannot reconstruct in detail the ecological situation of the past (Mayr 1970). This is a perfect example of how modern Darwinists "find shelter in ignorance", as

I. Popov, *Orthogenesis versus Darwinism*,
https://doi.org/10.1007/978-3-319-95144-7_9

Lyubishchev put it. We have to admit that extinctions, however intensively they are studied, often remain enigmatic.

Studies of some extinctions have been intensive indeed due to their applied significance: a geological layer can be characterised by its inhabitants, and the knowledge of changes in the fossil fauna facilitates the search for mineral deposits. Large-scale international research projects are devoted to extinctions marking major "biostratigraphic events" or "bio-events" but in several reviews on this topic species senescence is not mentioned at all. Moreover, paleontologists tend to ignore biological factors altogether. Some authors express an extremely ectogenetic view on evolution: were it not for the stimulating impulses of environment, life would have stopped at the level of protobionts (Krasilov 1984). Evolving organisms are considered as an amorphous mass, which changes only under an external influence. In anthropological studies such a concept would signify that without environmental changes humans would not grow but would remain in the condition of a child for an indefinite time.

Most explanations of extinction mechanisms focus on abiotic factors. Numerous hypothesis have been suggested in this connection such as: (1) The revolution of the solar system around the centre of the Galaxy results in changes of conditions around the Earth (cosmic rays, magnetic fields etc.), which negatively affect the biosphere and result in mass extinctions. (2) Large meteorites falling on the Earth cause natural disasters. (3) The explosions of the Supernova influenced the biosphere killing almost all living organisms. (4) Periodic surges in volcanic activity cause natural disasters and negative changes in the biosphere. (4) Sudden orogenic processes changed climate and other environment characters. (5) Oceanic inversion: changes of the zones of up- and down-welling resulted in climate changes and caused mass extinctions. (6) Sudden changes in air transparency caused a general change of vegetation and thus of the animal world. (7) Changes in the gas composition of the atmosphere, in particular the oxygen level, caused mass extinctions because organisms did not have time to adapt to new conditions. (8) Bursts of solar activity had a negative impact on the biosphere. (9) The Ocean level changed causing sudden changes of the environment. (10) The Moon emitted dust shadowing the Earth and suppressing vegetation and the entire biosphere. (11) Photoperiodic patterns changed because of some cosmic processes, and some organisms could not adapt to new conditions. (12) There is an additional planet called Nemesis in the Solar system; it has a very large orbit and approaches the Earth once per several millions years causing natural disasters (Walliser 1986, 1996; Nesov 1991; Alekseev 1998).

Biotic factors have rarely been analysed. If they are taken into account, the primary action of new organisms is usually postulated: they evolve and exterminate their predecessors. In modern studies a more complex scheme is usually suggested: the displacement of a species by another could be a catastrophe for species ecologically linked with the "old" one. For example, as flowering plants substituted coniferous plants, they automatically exterminated animals associated with the latter. Such a displacement can "shake ecological pyramids", cause "crisis" and, eventually, a global change of flora and fauna (Zherikhin 2003).

These hypotheses often border on science fiction. Nevertheless, their number is growing steadily though they are always either insufficient or unconvincing. The greatest unclearness is associated with the fact that extinction of species from certain groups of high taxonomic level was selective. At the same time, extinct and persisted species could not differ in their capacity to resist negative impacts. Why did all ammonoids die out in the end of the Mesozoic while several nautiloids, persisted? Why did dinosaurs became extinct while crocodiles and turtles live happily nowadays? Why did rudists become extinct, while oysters persisted? Why did bony fishes force out many cartilaginous fishes but not sharks and rays? Why, if mammals were downfall of dinosaurs because they ate their eggs, did they not similarly consume the eggs of turtles and crocodiles? There is a long list of such questions. Several scientists suggest that it was simply a matter of luck (Clarkson 1998). However, surviving species did have several "non-casual" characters. When extinct species flourished, they were primitive, non-variable and not numerous. It would seem that after negative changes of the environment or the appearance of competitors at least some representatives of broadly distributed dominant groups (e.g. ammonoideans) should have survived. However, such groups became totally extinct while their more primitive and less numerous relatives (e.g. nautiloids) survived.

Further existence of organisms that succumbed to extinction might have been prevented by some self-evolutionary processes. Such an interpretation agrees with the fact that the extinctions mentioned above were slow and gradual and occurred over millions of years. Curt Teichert, writing about the Permian extinctions, compared this process to the Farewell symphony of Haydn: "The way in which many Paleozoic life forms disappeared towards the end of the Permian Period brings to mind Joseph Haydn's Farewell Symphony where, during the last movement, one musician after the other takes his instrument and leaves the stage until, at the end, none is left." (Teichert 1990, p. 231). The idea was echoed by Desmond Donovan in connection with the extinction of ammonites: "[…] all ammonitologists now must know that the decline in diversity in the Upper Cretaceous was gradual over many million years. The extinction wasn't a sudden event—it was more like the end of Haydn symphony in which the players creep out by one." (Donovan 1993, p. 7). Meteorites or other similar impacts probably killed the last "performer", who was about to leave anyway.

Difficulties associated with the explanation of bio-events are less visible in the most synthetic concept on the topic, that by **Otto H. Walliser** (1996). His main idea is as follows: changes of the biosphere have been numerous and manifold and so were their causes; different impacts could cause similar events and, on the contrary, similar impacts could cause different events. The history of the Earth is a complex tangle of circumstances, and in different periods different factors were of major significance. Thus, at the border of the pre-Cambrian and the Cambrian the change of the biosphere was caused by a "biological innovation", the origin of skeletons and other features, which resulted in the "Cambrian explosion" of new phyla. Later, climate changes associated with glaciations caused the change of the biosphere at the Ordovician-Silurian boundary. In the middle of the Devonian the Ocean level changed (as indicated by the black shale deposits), and the next significant change, the Kellwasser event, occurred. At the end of the Palaeozoic, extinctions were a culmination of long climate changes related with the changes in the Ocean level. Intensive Jurassic evolution of ammonoideans is an example of an event determined by orogenic factors. Finally, the mass extinction at the end of the Mesozoic is a remarkable example of extinction caused by cosmic impact.

The latter interpretation seems especially improbable in the light of the above discussion. Nevertheless, this hypothesis is now quite popular because of the recent discovery of the iridium anomaly: high levels of iridium at the borderline between the Mesozoic and the Cainozoic. Such levels are usually thought to indicate a meteoritic impact. The coincidence of the iridium anomaly and the final extinction of many groups gave rise to blood-curdling stories about a world-wide catastrophe, when the sky remained dark for years because of dust and almost the entire biosphere perished. The iridium anomaly was communicated in *Science* and other prestigious journals (Alvarez et al. 1980), which is probably why it is not being ignored although it is quite impossible that meteorites could act selectively and gradually on certain groups of organisms. The authors of this discovery did not explain this, focusing mainly on the inorganic aspect of the anomaly.

It is interesting that despite the diversity of mechanisms of bio-events, they were shown to have a certain periodicity. There are some indications that mass extinctions and mass originations occurred in turns in equal periods of time. Unfortunately, there is no consensus about the duration of these periods: 8, 12, 26, 32 millions of years were given. There is no convincing explanation of mechanisms that could underlie this biological periodicity. By exclusion, the authors of the most authoritative research in this field concluded that these mechanisms must have a cosmic nature (Raup and Sepkoski 1984) because some cosmic phenomena such as planet revolutions are periodic. No other explanation has been suggested. This is an acknowledge of complete defeat: an appeal to unknown cosmic factors which affected biosphere by an unknown mechanism in an unknown period of time is hardly different from theological explanations of natural phenomena. Modern interpretations of bio-events are reminiscent of ancient beliefs, where nature and human activity were thought to result from the impact of heavenly, underground and underwater forces. In Greek mythology Zeus, the god residing in heaven, Poseidon, the god of water, and Hades, the god of the underground world, were the main authority, with Zeus being the highest one. Likewise, in modern research of the biosphere evolution "the thunder from heaven", i.e. meteorites, stand a little higher in the hierarchy of factors than volcanoes and ocean cataclysms but generally all these "gods" are considered more powerful than the action of biosphere.

As mentioned above, several orthogeneticists pointed out evolutionary regularities such as acceleration, retardation, specialisation etc. shared by very different groups of organisms. The "law of Rosa" (the law of decreasing variability) could be a good characteristics of species senescence. If variability really drops down to zero in the course of evolution, this could explain "phylogenetic ageing" (Rosa, 1931). However, this law has never been proven, nor was it suggested how variability could be measured. The classics of modern Darwinism sometimes also claimed that variability decreased with evolution. This would seem reminiscent of Rosa's law but

Darwinists actually meant not the potential but the realised variability: since species gradually become increasingly more adapted to a certain environment, at some moment there is no reason to change anymore. That is why the variability of an "ideally" adapted organism is close to zero (Schmalhausen 1969).

The possibility of ageing or "tiredness" of taxa is often mentioned as something self-evident in studies, especially paleontological ones, which do not deal with general evolutionary concepts directly. A sign of ageing and a symptom of imminent extinction is the appearance of "extravagant", "grotesque", "overspecialised", "excessive", "abnormal", "pathological", "disharmonious", "decentralised" or "deviant" types. The list of such synonyms can be continued since a common definition of these types has never been found.

In some groups these "extravagances" are manifested as spines. An American palaeontologist **Charles Emerson Beecher** (1956–1904) discussed the phenomenon of spinosity in a paper entitled *The Origin and Significance of Spines* (Beecher 1898; see Davitashvili 1948 and Bowler 1983). According to Beecher, spines appear at later stages of a group's evolution, then increase in size and number and soon after the group becomes extinct. This process, discovered in studies of trilobites and brachiopods, is similar to the individual development of these animals. A sleek rounded embryo or larva acquires increasingly more differentiated characters including spines, which reach their maximum development in old individuals. In the same way, in evolution the first representatives of the group are spineless and then gradually develop increasingly more complex spines. They betray the senility of the group, which soon becomes extinct or persists as a handful of primitive representatives without highly developed spines (Fig. 9.1).

Spinosity is a common condition in animals, plants and protozoa. Spines are traditionally regarded as protective devices but a close examination shows that this

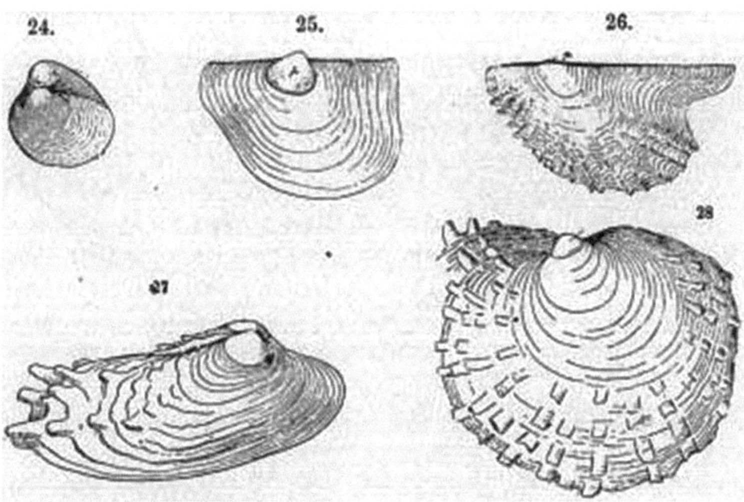

Fig. 9.1 Spines in young and senile species of brachiopods (Beecher 1898)

is not always so. Beecher concluded that spines evolve regardless of the benefit for the owner as a manifestation of a certain evolutionary stage (Beecher 1898).

Indicators of phylogenetic senescence were found in several other groups. **William Dickson Lang** (1878–1966) noted them in bryozoans: the accumulation of calcium becomes excessive in their evolution, leading to the formation of skeletal outgrowths and spikes (Taylor 2002). The complexity of shell shape increases gradually in the evolution of cephalopods. Large horns of dinosaurs and mammals are also such indicators. The disappearance of teeth in vertebrates is a harbinger of a rapid demise (Davitashvili 1948). In reptiles, a large size and outgrowths on various parts of the body (crests on the back, proliferation of parietal and frontal bones) are also considered as a sign of phylogenetic ageing (Kaiser 1970). Such cases were noted to be similar with an unhealthy condition of "normal" or "young" types. Sometimes, after the emergence of extravagant types a short "rejuvenation"—re-emergence of characters typical of the early evolutionary stages—occurs. In a way, the group "becomes childish" before its demise.

There is a rich collection of "foolish" creatures in the fossil record, and some paleontologists got quite carried away speculating about their lack of vitality. The description of *Diplodocus* by **Hans Elmar Kaiser** is a good example (Kaiser 1970). Kaiser reconstructed the size of its various organs such as lungs, heart and brain and found that its heart was very small in respect to the body. Kaiser compared *Diplodocus* with an overgrown man with a weak heart. He thought that veins and capillaries just could not function at a distance of several meters from the heart because of Torricelli's vacuum. The nervous, the respiratory and other systems of *Diplodocus* seemed to be so ridiculously arranged that it was unclear how it could be alive at all (Kaiser 1970). It was considered as a clear evidence of "decentralisation" causing and age-related demise of a phylogenetic trend.

These speculations are interesting and to some extent convincing. However, they are hard to reconcile with the fact that *Diplodocus* existed for a long time in large numbers (otherwise, modern researchers would not have so many of their fossil remains at their disposal). Moreover, Diplodocus was not the only dinosaur with a seemingly "absurd" design. Other researchers might well consider *Diplodocus* and their relatives as normal animals without any pathology. Kaiser noted that species and organisms are not disharmonious themselves but in comparison with related groups or the "type" (*Bauplan*). More detailed information about types and deviations from them is difficult to find. There is a tautology here, just as in the case of specialization: "extravagant types" are considered as a consequence of ageing but ageing itself is determined by the presence of such types. It is apparently more difficult to observe ageing in modern species but if some of them (e.g. giraffes) would disappear, future paleontologists might be tempted to explain this sad event in the same way as the extinction of *Diplodocus* is explained now.

"Extravagant" forms among modern animals were sometimes interpreted as a manifestation of an opposite phenomenon—a high variability. For example, a mammal called babirusa on the Sulawesi Island looks like a pig with strangely shaped tusks, which grow upwards while their tips are bent inwards. It is difficult to imagine how these tusks can be of use. They seem to be suitable only for fixing the head on tree branches, which babirusas apparently do not need. Rensch considered this as an evidence of unlimited variation: nature provides all possible material for selection, useful and not. In the case of the babirusa, a meaningless variant was accidentally preserved by selection because it was not harmful (Rensch 1954). Other similar cases have been explained by the action of sexual selection. Both these explanations, however, are barely convincing.

There are other indications of species senescence in modern organisms besides "extravagant" types. Many "ageing" species were found among animals living in caves. Depigmentation, slow development, "unhealthy" paleness were considered as indications of an imminent danger of extinction (Vandel 1965). However, other researchers of cave animals have not seen any defectiveness in them. Moreover, there were some indications that "normal" animals can originate from cave dwellers. A recent review of these studies concludes that speculations about the origin of cave fauna do not explain anything because they are based on statements that cannot be proven (Desutter-Grandcolas 1997).

A French writer **Henri Decugis** (1874–1947) cited numerous examples of species senescence in his book *Le Vieillissement du monde vivant* (*Senescence of the living world*) (Decugis 1943). He claimed that all extant taxa without exception demonstrated senile characters and were close to extinction. Whales, brachiopods, anteaters and armadillos, disk-shaped fish etc. were too large, or too archaic, or too flat, or too bizarre. Decugis believed that only humans could resist species senescence due to mentality. This viewpoint, though it might seem absurd, is difficult to challenge. It is known that many species became extinct, and it is logical to suppose that modern ones will be extinct too. The current stage of their existence can be considered as an evolution towards extinction, i. e. senescence. Of course, it is hard to estimate whether they have already started to age or not, and some scientists tend to speak ironically about this approach. The advocates of orthogenesis have been compared with a pessimist reading an annual list of obituaries and exclaiming: "What is the world coming to? Here are all these great men dead during the year and not a single great man born!" (Jepsen 1949, p. 493).

Though it is difficult to characterise species senescence and the origin of "excessive" types, it is evident that numerous extinctions over millions of years cannot be explained convincingly in terms of the natural selection theory. It seems that organisms evolve spontaneously until a certain limit and then become enable to give rise to new "adapted" forms or at least to conserve the existing ones. Irrespective of the attitude to orthogenesis, the existence of enigmatic processes of disharmony, decentralisation, overspecialisation, decrease of variability etc. is usually recognised. So, **Louis de Bonis** in a review of extinctions mentioned these processes in the list of the causes of extinctions even although he called the idea of species senescence "anthropomorphic" and "tautological" (Bonis 1991). He was against the use of the term "ageing" in such context, because it did not give information on the mechanism of extinctions. However, the mechanisms of the senescence of organism are not too well known either but no one would argue that this phenomenon exists. It is interesting that one of the numerous theories explaining the mechanisms of individual ageing points to the fact that the terminal parts of chromosomes of somatic cells become shorter in process of cell division. At some point cells cannot divide anymore, and some organs and tissues cannot function normally, hence the ageing (Olovnikov 1996). It may be that something "shortens" in the entire genome too, and it cannot be reproduced any longer. However, the telomere concept, similarly to many other concepts aiming to explain ageing, has been confirmed only partly. This is likely to mean that ageing results from numerous processes, and different processes may cause it in different organisms (Anisimov 2008). That is why only some processes related to senescence can be suppressed, and if they are, others will inevitably originate. For this reason, senescence cannot could be totally exterminated. The mechanisms of species senescence are likely to be numerous as well.

Thus, the claim on the existence of evolutionary changes caused by species senescence is rather solid. It attracts attention to the internal forces of evolution or at least indicates that the explanation of evolutionary events cannot be reduced to the action of natural selection on unlimited variability. Evolving groups can lose the ability to provide the material for selection because of their senescence, which results in the emergence of "unfit" organisms.

References

Alekseev AS (1998) Mass extinctions in the Phanerozoic (Massovye vymiraniya v fanerozoe). Extended abstract of Cand Sci (Geol) Dissertation. Moscow

Alvarez LW, Alvarez W, Asaro F, Michel HV (1980) Extraterrestrial cause for the cretaceous–tertiary extinction. Science 208(4448):1095–1108

Anisimov VN (2008) Molecular and physiological bases of ageing (Molecularniye i fiziologicheskiye osnovy stareniia). Nauka, St Petersburg

Beecher CK (1898) The origin and significance of spines: a study in evolution. Am. J. Sci., 4th series, 5(6): 1–20

de Bonis L (1991) Evolution and extinction in the animal kingdom (Évolution et extinction dans le règne animal). Masson, Paris

Bowler PJ (1983) The eclipse of Darwinism. Johns Hopkins Univ Press, Baltimore

Clarkson ENK (1998) Invertebrate paleontology and evolution. Blackwell Science Ltd

Davitashvili LS (1948) The history of evolutionar palaeontology from Darwin to our days (Istoriya evolyutsionnoy paleontologii ot Darvina do nashikh dney). AN SSSR, Moscow/Leningrad

Davitashvili LS (1969) The causes of extinction of organisms (Prichiny vymiraniya organizmov). Nauka, Moscow

Decugis H (1943) Ageing of the living world (Le viellissement du monde vivant). Masson et C-ie, Paris

Desutter-Grandcolas L (1997) Studies in cave life evolution. A rationale for future theoretical developments using phylogenetic inference. J Zool Syst and Evol Res 35(1):23–31

Donovan DT (1993) Ammonites in 1991. In: House MR (ed) The Ammonoidea: Environment, Ecology and Evolutionary Change. Systematics Association Special, vol 47. Clarendon Press, Oxford, pp 1–12

Jepsen GL (1949) Selection, "orthogenesis", and fossil record. Proc Am Philos Soc 93(6):479–501

Kaiser HE (1970) The abnormal in the evolution (das Abnorme in der evolution). E. J. Brill, Leiden

Krasilov VA (1984) The theory of evolution: a need for a new synthesis (Teoriya evolyutsii: neobkhodimost' novogo sinteza). In: Krasilov VA (ed) Evolutionary studies. Marcoevolution (Evolyutsionnye issledovaniya Makroevolyutsiya) DVNTs RAN, Vladivostok, pp 4–12

Mayr E (1970) Populations, species, and evolution. Belknap Press of Harvard University Press, Cambridge (Mass)

Nesov LA (1991) Pterosauri over plane-tree forests and brackish-water bays of seas (Krylatye yashchery nad platanovymi lesami i solonovatymi zalivami morey). Gerpetologicheskie issledovaniya 1:147–163

Olovnikov A (1996) Telomeres, telomerase, and aging: origin of the theory. Exp Gerontol 31(4):443–448

Raup DM, Sepkoski J (1984) Periodicity of extinctions in the geologic past. Proc Nat Acad Sci USA 81:801–805

Rensch B (1954) New problems of the evolutionary theory: transspecies evolution (Neuere Probleme der Abstammungslehre: transspeciphische evolution). Enke, Stuttgart

de Rosa D (1931) Ologenesis. A new theory of evolution and geographic distribution of living beings (L'Ologénèse. Nouvelle théorie de l'évolution et de la distribution géografique des êtres vivants). Libraire Félix Alcan, Paris

Schmalhausen II (1969) The problems of Darwinism (Problemy darvinisma). Nauka, Leningrad

Taylor PD, Lang WD (2002) orthogenesis and the evolution of Cretaceous cribrimorph bryozoans. In: Patrick N, Wyse J, MES J (eds) Annals of Bryozoology. International Bryozoology Association, Dublin, pp 275–299

Teichert C (1990) The Permian-Triassic boundary revisited. In: Kauffman EG, Walliser OH (eds) Extinction Events in Earth History. Springer, Berlin/Heidelberg/New York, pp 199–238

Vandel A (1965) Biospeleology. The biology of cavernicolous animals. Pergamon Press, Oxford

Walliser OH (1986) Towards a more critical approach to bioevents. In: Walliser O (ed) Global bioevents in earth history. Springer Verlag, Berlin, pp 5–16

Walliser O (1996) Patterns and causes of global events. In: Walliser O (ed) Global and event stratigraphy in the Phanerozoic. Springer, Berlin, pp 7–19

Zherikhin VV (2003) Selected works in palaeontology (Izbrannye trudy po paleoekologii). KMK, Moscow

Chapter 10
Adaptation or Non-adaptation?

Abstract The adherents of directed evolution claimed that adaptation to the environment was not the main driving force of evolution. Their opponents either denied instances of inadaptive evolution or insisted that adaptive changes were by far more numerous and therefore evolution in general was the creation of adapted organisms. Orthogeneticists objected that seemingly adaptive organs first evolved independently of the environment and then an organism put them to use. In this way, organisms adapted to their own structure. The phenomenon of senescence is a remarkable example of non-adaptive nature of characters.

Keywords Adaptation · Preadaptation · Panglossian paradigm · Inadaptive evolution · Senescence

The existence of non-adaptive evolutionary trends was another serious argument presented by orthogeneticists. They claimed that adaptation to the environment was not the main driving force of evolution and criticised the idea of adaptation on the basis of various material. Especially convincing arguments were based on fossil data.

Some palaeontological trends are indeed, to all appearances, non-adaptive. Interestingly, several Darwinists recognised this as well. Wladimir O. Kowalevsky, a passionate supporter of Darwin, eagerly looked for evidence of natural selection in his paleontological studies but found numerous "inadaptive" forms as well. In particular, he characterised adaptive and inadaptive reductions of the number of fingers in ungulates and inadaptive structure of teeth in some ungulates and mastodons.

Leo Davitashvili, a vocal critic of all opponents of natural selection theory, collected examples of non-adaptive evolution too. His list included the tail of pterosaurs, imperfect teeth of some predatory pelycosaurs, the structure of tree stems in ancient cryptogams with a centripetal growth of wood (clearly less favourable than centrifugal) etc. He thought that these observations could provide a base for broad generalisations and even suggested the "law of Wladimir Kowalevsky" or the law of adaptive and inadaptive evolution (Davitashvili 1948, p. 393).

I. Popov, *Orthogenesis versus Darwinism*,
https://doi.org/10.1007/978-3-319-95144-7_10

The admission of non-adaptive evolution may seem inconsistent with an active support of the natural selection theory. Davitashvili overcame this in the following way: the words "adaptive" and "inadaptive" are relative; the pattern of an organ's evolution is called adaptive if it has some advantages over another pattern occurring in a related form. This verbal trick did not fool anyone though and "the law of non-adaptive evolution" was not supported by other Darwinists.

The theory of natural selection focuses on adaptation, while maladaptive series of successive forms illustrate orthogenesis. Extinct cephalopods have often been mentioned in this regard. So, Edgar Dacqué described the series of nautiloids, in which the shift from straight shells to curved ones was observed. However, straight shells re-emerged in some groups though they were condemned to extinction by natural selection (cited in Sobolev 1924).

In the times after Dacqué, extinct cephalopods, especially nautiloids and ammonoideans, also provided material for various evolutionary heresies. These two groups have much in common: both had external shells subdivided into chambers, the chamber contacting the environment contained organs while other chambers were filled with gas ensuring buoyancy. A line marking the chambers on the surface of the shell (suture line) serves as a diagnostic character for the distinction of these groups: it is straight or slightly bent in nautiloids and sinuous in ammonoideans (Figs. 10.1 and 10.2). The suture line in the first ammonoideans was similar to that

Fig. 10.1 Representative of nautiloids *Nautilus inornatus*, Musée d'Histoire Naturelle, Paris

Fig. 10.2 Representative of ammonoids *Aulacostephanus yo*, Musée d'Histoire Naturelle, Paris

in nautiloids but then its complexity increased. Late ammonoideans (ammonites) bear fanciful ornamentations on the shell surface.

An increased complexity of the suture line (and, correspondingly, of the septa between chambers) is usually considered as an adaptation to life in deep waters. However, a more thorough analysis does not confirm this explanation. Nautiloids, which have a simple suture line, are well adapted to depth. Modern nautiloids were registered at a depth of up to 504 m and might live even deeper, at a depth of 800–850 m (Saunders and Ward 1987). Calculations of the animal's resistance to water pressure do not confirm the advantage of sinuous septa either (Hewit 1993).

There are other hypothesises explaining the benefits of a complex suture line but all of them are disproved by the fact that all ammonoideans became extinct while nautiloids, which seem "the least fittest", did not. Evolution of shell shape in general is also difficult to understand. At the final stage of ammonoids' evolution the so-called heteromorph ammonites, with irregularly coiled shells, often evolved. This baffles the researchers completely. "What influence, if any, environmental factors can have had in driving this evolution, in accordance with the classical Darwinian adaptive canon, remains wholly unknown. Numerous claims in the literature to the contrary are almost invariably tautological. In short, we now know rather well how the ammonites evolved, but not why" (Calloman 2002). "The shells of Calloman's Cardioceratidae got larger from time to time, they got smaller, they got fatter, they got thinner again, they acquired a cadiconic shell form, they lost it in due course. Were these changes adaptive responses to different environmental demands which then changed again, and led to a change of different direction? Were they not? If not, why did they happen at all? Some of the changes affected animals over a wide area of the globe, so the adaptation to local environmental changes is not a plausible explanation" (Donovan 1993).

Inadaptive evolution is sometimes illustrated by modern cases too. Some organisms just seem too complex. Orchids, for example, have very intricate flowers, which sometimes repel rather than attract insects. Pollination is extremely difficult because a certain species of insect is required. Orchid pollen is packed in such a way that it is completely pulled out after a single visit of an insect, and the risk of its loss is relatively high. Finally, in order to grow into a plant, seeds require certain fungi and other specific conditions. As a researcher of orchids remarked: "As though in some trends evolution proceeds in a chosen direction, following a blind inclination, ignoring any needs and not being able to pave any other ways" (Neger 1913 cited in Berg 1977, p. 176).

The opponents of orthogenesis have either denied facts of inadaptive evolution or claimed that adaptive changes are by far more numerous and therefore evolution in general is a creation of adapted organisms. Orthogeneticists have objected that seemingly adaptive organs evolve without any relation to the requirements of the environment and an organism finds an application for them, that is, adapts to its own structure. This idea is mainly known in the guise of the concept of preadaptation by Lucien Cuénot but it was also expressed by other authors. As early as in 1829 Étienne Geoffroy Saint-Hilaire warned about the danger of teleology or the "philosophy of final causes", which is greatly reminiscent of the modern adaptationistic paradigm. "... naturalists, undoubtedly without attaching too much importance to these explanations, find delight in explaining what they see by the necessary connections, of which they have shifted the starting point. In this abuse of the philosophy of final causes, the effect is made to produce the cause. So, on noticing a bird flying in the air, one concludes that it was given an organisation to suit this purpose; one adds further that it should have hollow bones in order to weigh less, an ample covering of light feathers, an extraordinary enlargement of its anterior extremities etc. In the same way, it has been said about fish that, since it lives in a more resistant milieu than air, its motor forces are calculated to afford it a certain mode of progression; that because it forms part of the branch of Vertebrates, it has to have an interior skeleton. And then, arriving at the logical conclusion of this reasoning, these naturalists seem enraptured by the fact that all these necessary things meet together simultaneously and a point where they are required in order to sanctify the principle of order and harmony manifested in all the products of creation.

Reasoning in this way, you may well say that a man who uses crutches was originally destined to the misfortune of having one of his legs paralysed or amputated. Let us remain the historians of what is; let us not conclude about functions before having established or having tried to establish the instruments that produce them. Every being came from the hands of the Creator with proper material conditions; it is capable of things according to what it was attributed to be capable of doing; it uses its organs according to their capacity to action" (Geoffroy Saint-Hilaire 1829, p. 8).

Morphological regularities are still being interpreted in this way. Statements of the kind mocked by Saint-Hilaire can be found in studies in evolutionary developmental biology (evo-devo): aquatic animals got into water first and started to adapt to it next; the ancestors of snakes started to adapt to crawling and in doing so lengthened and lost limbs (Bejder and Hall 2002). This is hardly different from teleology provided the words "purpose" or "final cause" are used often enough. Limbless snakes are adapted to crawling in the same way as a man who lost a leg is adapted to moving on crutches.

With the development of evolutionary biology, the opponents of Darwinism often protested against the "abuse of teleological philosophy". Their idea of pre-adaptation seems a natural reaction to a superficial interpretation of adaptation, in which cause and effect are confused. In this respect, it seems worthwhile to recall what Berg said about flying fish. According to Darwin, they adapted to flight. Under certain conditions, selection pressure favours characters allowing fish to make long jumps, in particular, long fins. They gradually became longer and longer, fish jumped further and farther, and finally modern flying fish evolved. A little more of the same, it would seem, and they would fly like birds. Berg did not accept this explanation. He argued that the fins of flying fish lengthened regardless of the requirements of the environment. There are fish with short fins that like to jump out of water as well as long-finned fishes that live near the bottom. Anyway, fish do not really need to fly, do they? An aquatic environment is better for them, and they may jump for no particular reason. Berg gave examples confirming that long fins are just one of the possible options, which is normally present in some species of fish or occurs as an anomaly in species with normal fins. This variant arises spontaneously, not in response to environmental requirements, though it may seem adaptive under certain conditions (Berg 1977).

Cases of seeming adaptation crop up in some studies irrespectively of collisions of Darwinism and orthogenesis. For instance, one of evolutionary trends of salamanders is adaptation to arboreal lifestyle, that is, moving on flat, slippery and wet surface of leaves of tropical trees. This adaptation is achieved by a decrease of size, an acquisition of webs between fingers and a partial reduction of phalanxes. In general, the degree of the development of webs correlates with the degree of "arboreality". However, there are exceptions: non-arboreal salamanders with developed webs (*Bolitoglossa dafleini*) and "acrobatic" arboreal salamanders (*Bolitoglossa franklini*) with very small webs (Alberch 1981, p. 95). In the light of these exceptions, the viewpoint of orthogeneticists about preadaptation seems reasonable enough.

Entirely neutral evolutionary trends, having nothing to do with adaptation to either the environment or the organisms' own structure, are known. One such trend was described during an analysis of the diversity of passerine birds. "At the same time, there are a number of features that are not related, at least apparently, with the external environment. [...] In the group of passerine birds, which is extremely rich in genera and species, some obscure physiological process invariably leads, with time, many species, sometimes quite distant taxonomically, to red colouration. This happens in the same way in different locations, under different environmental conditions and in very different climates. In the metaphoric phrase of academician Sushkin, red colouration of finches is a "general's uniform" to which some species get promoted. It is usually preceded by yellow colouration, while yellow is preceded by brown. We do not know yet what the matter is from physiological viewpoint. The climate can only partially delay or slightly modify these outfits but nothing more." (Serebrovsky 1928, p. 128).

Thus, some organisms are badly adapted, others have adapted "by chance" rather than in response to the demands of the environment and others still have acquired characters indifferent in respect of "adaptation". This is, roughly, the conclusion one gets from the "orthogenetic" analysis of adaptation. In other words, evolution may be adaptive, non-adaptive, preadaptive and neutral. Darwinists stress the adaptive way, while orthogeneticists pay more attention to the non-adaptive one. Who is right? Darwinists say that an inadaptive way is impractical and unnatural, and therefore cannot be the main trend. Anti-Darwinists claim otherwise. They have often drawn attention to the fact that the characters distinguishing large taxa are maladaptive such as a regular or an opposite arrangement of leaves, pentamerous or tetramerous flowers etc. (Nägeli 1884; Willis 1940). Who will be so bold as to try to prove the adaptive significance of these features?

Here, however, "there is a loophole where one can hide", as **Sergei S. Chetverikov** (1880–1959) put it (cited in Takhtadzhyan 1991, p. 501). This loophole is correlative variability. A maladaptive character may be linked in the heredity material with an adaptive one. For instance, a pentamerous structure of the flower might have been correlated in ancient plants with some important adaptive feature. However, even Chetverikov, one of the founders of population genetics and a forerunner of evolutionary synthesis, considered this explanation non-convincing. To accept it means to construct a theory on an extremely weak foundation: an unknown effect of unknown pleiotropic genes of unknown ancestors at an unknown time in the course of adaptation to unknown conditions.

The desire to explain biological phenomena from an adaptive viewpoint, whatever it takes, reduces biology to teleology. Gould and Lewontin (1979) aptly termed this approach "Panglossian paradigm" after the name of Professor Pangloss, a character in Voltaire's *Candide: or, Optimism.* This novel is a powerful attack on religious obscurantism. An omniscient philosopher, Professor Pangloss, can see the benefit in any of the terrible events befalling on the characters since he firmly believes in his teleological concept. He is extremely skilful in speculations about the usefulness of all things. "... noses were made to support glasses, and so we have glasses; two legs were obviously instituted to put shoes on, and so we have shoes. Stones were moulded to be cut and used to build castles" (Voltaire 1759 cited in Voltaire, 1887, p. 145). A terrible disease, a death by hanging, slavery in the galleys may be represented as useful for somebody or something. Finally, for the universal divine harmony "particular misfortunes make up common good; so that the more particular misfortunes, the better everything on the whole" (Voltaire 1759 cited in Voltaire, 1887, p. 145).

It seems that some people feel the need to admire everything that is going on and to prove to the others as well as themselves that everything is going in the right direction. (Perhaps, this is a suitable subject for psychology rather than evolutionary biology.) If this need is introduced into scientific biological research, a vicious circle arises: a researcher starts with from the idea that everything in evolution is useful, carries out a study and concludes that a process or phenomenon is useful indeed. Endless "new proof" of things long "proven" are popular in this field. Papers about "the leading role of selection" are still being published (Artamonova and Makhrov 2006). Numerous such publications have appeared in connection with the celebrations of the 150th anniversary of Darwin's *Origin of species* (Darwin 2009).

Panglossian paradigm is a waste of efforts and resources. Researchers have spent a lot of time measuring of the wings of *Drosophila* to reveal an adaptive significance of the smallest differences in the details of vein structure and concluded that they indeed have some adaptive significance although, they remark, these study are still at an early stage (Gilchrist et al. 2000)—and that in spite of the fact that thousands of wing measurements have been made. It seems that the researchers were fairly sure about the adaptive significance of differences in wing venation from the start. Why spend so much effort if the result is known in advance?

Measuring wings of fruit flies may seem an innocent occupation but fascination with this methodology may have serious consequences. This is shown by research on ageing. For obvious reasons, it is quite intense, and often based on a dogmatic view that to identify the mechanisms of a biological phenomenon one has to understand what selection pressure caused it. This means that researchers try to find benefits of ageing and associated diseases such as cancer and diabetes (Makrushin 2004). Common sense indicates that there can be nothing good in cancer and ageing. The lack thereof gives rise to ideas that ageing can be cancelled altogether. If it is a useful character that was promoted by natural selection, it can be replaced by non-ageing under a different selection pressure. Large-scale projects based on this idea are in progress and much money is spent under the slogan of cancelling senescence.

The authors of such studies refer to the classics of Darwinism. Indeed, the tradition of searching for the benefits of ageing dates back to Darwin's time. Some of Darwin's followers were inspired by arguments about the benefits of different variants of aging and death of representatives of a species. Alfred Wallace believed that since parents at some moment start competing for resources with their offspring, it is useful for a species if parents die shortly after reproduction. August Weismann thought that an accelerated aging was especially useful for a species, expressing this view in a series of lectures in the 1880s. He also believed that death was a kind of adaptation and that it was not inevitable. "I have endeavoured to show that the limitation of life in single individuals by death is not, as has been hitherto assumed, an inevitable phenomenon, essential to the very nature of life itself; but that it is an adaptation which first appeared when, in consequence of a certain complexity of structure, an unending life became disadvantageous to the species" (Weismann 1881, p. 111). Similar views are promoted a Russian academician Vladimir Skulachev, the author of a project aiming at eliminating ageing by medical means (Skulachev 1997, 2009).

Weismann's concept is hardly compatible with the fact that most organisms in nature do not live up to an old age. They usually perish earlier. Therefore ageing cannot be exposed to the action of natural selection. To remove this contradiction, it was suggested that there are genes of ageing, which have long been accumulating in the course of evolution but not manifesting themselves because their owners died early (Medawar 1952) or genes simultaneously responsible for ageing and some benefits for young organisms (Williams 1957). However, the existence of such genes has never been proven. On the other hand, genes involved in ageing were found in the simplest eukaryotic organisms, in which ageing is weakly expressed (Clark 2004).

To prove the benefits of ageing for species, examples from zoology and, less commonly, from botany have been used. It would seem that ageing, including accelerating ageing, is useful for some species (and so they do age) but not useful for others (and they do not age). A popular example from the first group concerns salmonid fish, especially the pink salmon, *Onchorhynchus gorbusha*. During spawning, these fish undergo a series of physiological and morphological changes, so that they inevitably die after spawning even if the environment is favourable. So, pink salmon undergoes an accelerated aging. It is believed that its biological significance is that dead adult fish fertilise the rivers where their offspring live. Various small organisms settling on dead salmons serve as food for salmon juveniles. Otherwise, cold northern streams would not have provided enough nutrients for them.

This example fits the dogma about the benefit of ageing so well that none of the researchers of ageing, it would seem, tried to examine it in detail. However, this example crumbles under close scrutiny. Eggs of pink salmon lie in the ground during autumn and winter, then the fries hatch and soon (during spring floods) begin to migrate rapidly downstream to the sea. At this stage they resemble a match (length 30 mm, weight 200 mg). Most such juveniles have empty stomachs during migration, that is, live mainly at the expense of the yolk sac. Even in the tributaries of the Amur—in the southern part of the distribution area, where biological productivity is relatively high—such "starving" individuals may compose up to 66.3%, 80% or 91.7% of the total number of migrating individuals (Levanidov, Levanidova, Levanidov and Levanidova 1957). In rare cases, when suitable habitats are formed owing to a certain configuration of rivers and lakes, pink salmon fries may delay in freshwater for a short time (until July) and feed a little there (Robins et al. 2005). A similar impracticality can be seen in the life cycle of chum salmon *Oncorhynchus keta*.

Among other salmons there are indeed cases when fertilisation of their freshwater habitats makes sense. Juveniles of the coho salmon (*Oncorhynchus kisutch*) migrate into the sea when they are two, three or four years old, those of cherry salmon (*Oncorhynchus masou*) migrate as two-year-olds, those of sockeye salmon (*Oncorhynchus nerka*) are aged one, two or three years when they migrate and juveniles of chinook salmon (*Oncorhynchus tschawytscha*) are aged one or two years. However, fertilisation of rivers probably benefits only a part of their populations. Firstly, the juveniles of these salmons actively feed on the fries of the pink salmon. Secondly, not all of them live in the far north, where waters are very poor in organic matter. They occur up to the southern border of the temperate zone (Korea, Hokkaido, California). In fact, excessive organic matter in the southern rivers can damage salmons because they need clean oxygen-rich water and pure ground for spawning. Most of them spawn in the river rapids, where the fertilizer is washed away quickly. Fertilizers may contribute to a better growth of juvenile salmon (especially sockeye salmon), if only some lakes of some lake-and-rivers systems are fertilised.

Atlantic salmon, whose habitats are similar to those of Pacific salmon, may die after spawning but it may also survive and return to the sea and then (next year) back to the river for the next spawning. In hatcheries, Atlantic salmons can be reared and repeatedly used for reproduction for many years. Under conditions of aquaculture, Atlantic salmon have been known to produce eggs for 10 years reaching the age of 14 years (Bilak 1998). Atlantic salmon leaves spawning grounds soon after spawning, so if they die after that, they die elsewhere.

The analysis of other salmonids aggravates the idea about the benefit of rapid senescence even further. So, the most "cold-water" salmonid species, the Arctic char (genus *Salvelinus*), does not "commit suicide" to provide its offspring with food. On the contrary, these fish live as long as they can, sometimes practicing cannibalism. In some northern lakes, the only food for large chars are their younger relatives. Thus, death after spawning is beneficial for the species only for some salmons. In fact, salmon species can be arranged in an order of ascending degree of manifestation of accelerated aging: (1) brown trout (*Salmo trutta*), chars, lenoks (*Brachymystax*), rainbow trout (*Onchorhynchus mykiss*), taimens (*Hucho*); (2) Atlantic salmon; (3) cherry salmon, sockeye salmon, chinook salmon, coho salmon; (4) chum salmon, pink salmon. In the first group there is no irreversible "suicide" after reproduction, in the second group "self-destruction" is frequent but not obligatory, in the third group "self-destruction" is a rule but there are exceptions (a tendency to form resident freshwater forms). In the fourth group adult fish always migrate to the sea, and dramatic death after spawning is mandatory. To note, all these species may occur under similar conditions, sometimes even in the same water bodies.

Many other animals also "commit suicide" after breeding but this phenomenon is unlikely to have any special adaptive significance. For example, squids die after reproduction (like most of pacific salmons) but their dead bodies "fertilize" the entire ocean where squid larvae are just a minute part of the immense mass of plankton. The same is true of polychaetes, chaetognaths and many other animals. In general, instances of programmed death after breeding are not a convincing argument of the benefit of ageing (Popov 2012).

Even less convincing are cases when organisms seemingly do not age because aging is unnecessary. Bats, which are notoriously long-lived, are believed to be such animals. They violate the "rule" generally valid for mammals and many other animals: maximum lifespan correlates the size, in other words, larger animals live longer than smaller ones. Bats live many times longer than other small mammals. Banding data suggest that bats may live up to 20–30 years. At the same time, they are among the smallest mammals: even relatively large bats with a wingspan of 30–40 cm may weigh as little as 10–20 g, while the smallest bats weigh 1–2 g—a record for mammals (Goswami 1999). The longevity of bats is usually regarded as a proof of the concept of aging benefits and a famous hypothesis about the accumulation of mutations. It is believed that bats "do not age" or age slowly because they have no enemies. The genes of aging were eliminated by selection from their ancestors as the latter did not die young. If animals have no enemies, the effect of selection weakens, they do not need to improve themselves in process of the struggle for existence and they do not need to evolve rapidly. In short, they can afford longevity (Skulachev 2009). The connection of the absence of enemies with slow aging also conforms to the hypothesis of pleiotropy of genes responsible for aging. Since they still have an opportunity to manifest, they are exposed to natural selection and can be eliminated in spite of their link to useful characters.

A review on bat longevity in connection with general problems of gerontology has recently been published (Brunet-Rossinni and Austad 2004). The longest living bat was Brandt's bat, *Myotis brandtii*, whose age in Siberia reached 38 years (Khritankov and Ovodov 2001). Bats of temperate zones spend several months in hibernation, and it was suggested that this is the reason of their longevity. However, tropical bats also live long, though not as long as their northern relatives. Though the longevity of bats still remains, by and large, unexplained, the researchers tend to agree that it is associated with the evolutionary concepts of aging mentioned above.

The authors of the review, however, leave out an important question: *how often* bats in natural environment reach an age of 20–30 years and whether they really have no natural enemies. Zoological studies indicate that bats have quite many enemies and long-livers are rare among them. Moreover, mammals such as skunks, moles, shrews and hedgehogs have relatively few enemies but are short-lived anyway. Complex schemes of selection of hypothetical genes seem to have a flimsy foundation (Popov and Kovalev 2010).

There is a simple explanation for the differences in the manifestation of ageing in animals mentioned above. They differ by the patterns of changes occurring during their lives. Salmon fish experience many changes, while the bats sleep most of their lives in a stable environment and often in absolute darkness. Light is a source of changes in an organism too, and accelerates ageing. Mice exposed to light for longer periods of time age more rapidly and die sooner (Anisimov 2007). Reproduction is another source of the changes of the body. Bats reproduce rarely (usually once a year), while other small mammals reproduce much more often. The greater the change, the greater the manifestations of aging, and vice versa.

These ideas might seem trivial yet they are a voice crying in the wilderness. An evolutionary hypothesis based on the search of selection pressures responsible for ageing is broadly used for justification of research and development programs. It turns out that the search for a mythical benefit of any biological process is not only false but also harmful. The realisation of this fact is bound to crop up in the scientific community, and with it, the concepts of directed evolution.

References

Alberch P (1981) Convergence and parallelism in foot morphology in the Neotropical salamander genus Bolithoglossa. 1. Function. Evolution 35(1):8–23

Anisimov VN (2007) Melatonin. The role in the organism, the use in medicine (Melatonin. Rol' v organizme, primenenie v meditsine). Sistema, St Petersburg

Artamonova VS, Makhrov AA (2006) Uncontrolled genetic processes in artificially sustained populations: proof of the leading role of selection in evolution (Nekontroliruemye geneticheskie protsessy v iskusstvenno podderzhivaemykh populyatsiyakh: dokazatel'stvo vedushchey roli otbora v evolyutsii). Genetika 42(3):310–324

Bejder L, Hall BK (2002) Limbs in whales and limblessness in other vertebrates: mechanisms of evolutionary and developmental transformation and loss. Evol Dev 4(6):445–458

Berg LS (1977) Works on the theory of evolution (Trudy po teorii evolyutsii). Nauka, Moscow

Bilak AT (1998) State of activitites on the recovery of Atlantic salmon stocks in Canada (Sostoyanie rabot po vosstanovleniyu zapasov atlanticheskogo lososya v Kanade). In: Kazakov RV (ed) Atlantic salmon (Atlanticheskiy losos'). Nauka, St Petersburg, pp 308–335

Brunet-Rossinni AK, Austad SN (2004) Ageing studies on bats: a review. Biogerontology 5:211–222

Calloman J (2002) Jurassic ammonites: real phylogeny in real time. In: The Lyell meeting: approaches to reconstructing phylogeny. The Geological Society, Burlington House, p 7

Clark WR (2004) Reflections on an unsolved problem of biology: the evolution of senescence and death. Adv Gerontol 14:7–20

Darwin (2009) Commemorations. http://darwin-online.org.uk/2009.html

Davitashvili LSh (1948) The history of evolutionar palaeontology from Darwin to our days (Istoriya evolyutsionnoy paleontologii ot Darvina do nashikh dney). AN SSSR, Moscow, Leningrad

Donovan DT (1993) Ammonites in 1991. In: House MR (ed) The Ammonoidea: environment, ecology and evolutionary change. Systematics association special, vol 47. Clarendon Press, Oxford, pp 1–12

Geoffroy Saint-Hilaire E (1829) A course of the natural history of mammals (Cours de l'histoire naturelle des mammifères). Pichon et Didier, Paris

Gilchrist S, Azevedo RBR, Partridge L, O'Higgins P (2000) Adaptation and constraint in the evolution of Drosophila melanogaster wing shape. Evol Dev 2(2):114

Goswami A (1999) Craseonycteris thonglongyai (on-line). Animal diversity web (Accessed June 29, 2009 at http:animaldiversity.ummz.umich.edu/site)

Gould SJ, Lewontin RC (1979) The spandrels of san Marco and the Panglossian paradigm: a critique of the adaptationist program. Proc R Soc Lond B 205:581–598

Hewit RA (1993) Relation of shell strength to evolution in the Ammonoidea. In: House MR (ed) The Ammonoidea: environment, ecology and evolutionary change. Systematics association special, vol 47. Clarendon Press, Oxford, pp 35–56

Khritankov AM, Ovodov ND (2001) Concerning longevity of Brandt's bats (Myotis brandtii Eversmann) in middle Siberia (O dolgozhitel'stve nochnits Brandta [Myotis brandtii Eversmann] v Sredney Sibiri). Plecotus et al 4: 20–24

Levanidov VY, Levanidova IM (1957) Feeding of the young of the summer chum salmon and the pink salmon in tributaries of the Amur (Pitanie molodi letney kety i gorbushi v pritokakh

Amura). Izvestiya Tikhookeanskogo nauchno-issledovatel'skogo instituta rybnogo khozyaystva i okeanografii 45:3–16

Makrushin AV (2004) Evolutionary predecessors of oncogenesis and senile involution (Evolyutsionnye predshestvenniki onkogeneza i starcheskoi involyutsii). Uspekhi gerontologii 13:32–43

Medawar PB (1952) An unsolved problem of biology. Lewis, London

NägeliC(1884)Mechanistic-physiologicaltheoryofevolutionarytheory(Mechanischphysiologische Theorie der Abstammungslehre). Druck und Verlag von R. Oldenbourg, Munich and Leipzig

Neger FW (1913) Biology of plants on experimental basis (bionomics) (Biologie der Pflanzen auf experimentelle Grundlage [Bionomie]). Stuttgart cited in berg LS (1977) works on the theory of evolution (Trudy po teorii evolyutsii). Nauka, Moscow

Popov IY (2012) Distribution of various aging patterns in the system of the animal world. Adv Gerontol 2(1):1–9

Popov Iyu, Kovalev DN (2010) Obstoyatel'stva dolgozhitel'stva letuchikh myshey (v svyazi s materialami issledovaniy na territorii Sankt-Peterburga i Leningradskoy oblasti) (The circumstances of lonevity of bats) [in connection with the materials of studies in St Petersburg and the Leningrad Region]. Uspekhi gerontologii 23(2):179–185

Robins JB, Abrey CA, Quinn TP, Rogers DE (2005) Lacustrine growth of juvenile pink salmon and a comparison with sympatric sockeye salmon. J Fish Biol 66:1671–1680

Saunders WB, Ward PD (1987) Ecology, Distribution and Population Characteristics of Nautilus. In: Saunders WB, Landman NH (eds) Nautilus: The Biology and Paleobiology of a Living Fossil. Plenum press, New York, pp 137–162

Serebrovsky PV (1928) Darwinism and the teaching about orthogenesis (Darvinizm i uchenie ob ortogeneze). In: Kozo-Polyansky BM (ed) The theory of nomogenesis. A new phase in the development of Russian anti-Darwinism. A collection of critical papers (Teoriya nomogeneza. Novaya faza v razvitii rossiyskogo antidarvinizma. Sbornik kriticheskikh statey). Izdatel'stvo Gosudarstvennogo timirhyazevskogo nauchno-issledovatel'skogo instituta, Moscow, pp 87–158

Skulachev VP (1997) Ageing of an organism is a special biological function, not the result of a breakdown of an complex living system: a biochemical substantiation of Weismann's hypothesis (Starenie organizma – osobaya biologicheskaya funktsiya, a ne rezul'tat polomki slozhnoy zhivoy sistemy: biokhimicheskoe obosnovanie gipotezy Veysmana). Biokhimiya 62(11):1394–1399

Skulachev VP (2009) How to cancel the programme of ageing of the organism? (Kak otmenit' programmu stareniya organizma?) Rossiiskiy khimicheskiy zhurnal 53: 125–140

Sobolev DN (1924) The fundamentals of historical biogenetics (Nachala istoricheskoy biogenetiki). Gosudarstvennoe izdatel'stvo Ukrainy, Simferopol

Takhtadzhyan AL (1991) Darwin and the modern theory of evolution (Darvin i sovremennaya teoriya evolyutsii). In: Russian edition: Darwin Ch (1991) The origin of species (Proiskhozhdenie vidov). Nauka, St Petersburg, pp 489–523

Voltaire (1759) Candide, ou l'Optimisme. Cited in Voltaire (1877) Candide, ou l'optimisme. In: Œuvres complètes de Voltaire, 21, pp 175–179, Garnier

Weismann A (1881) On the duration of life (Ueber die Dauer des Lebens). In: Tageblatt der Versammlung deutscher Naturforscher und Aerzte in Salzburg. Salzburg Jg, vol 54, pp 98–114

Williams GC (1957) Pleiotropy, natural selection and the evolution of senescence. Evolution 11:398–411

Willis JC (1940) The course of evolution by differentiation or divergent mutation rather than by selection. Cambridge University Press, Cambridge

Chapter 11
Anticipation, Parallelisms and Convergences

Abstract Phylogenetic anticipation means that organs or their rudiments evolve at an early stage of a group's evolution, disappear at the next stages and later re-emerge either as adaptations or as a mark of a new evolutionary stage. A similar phenomenon is the emergence of aristogenes, organs useless for an organism but creating preconditions for the origin of "useful" organs in its distant descendants. Unlike "anticipatory" features, they do not disappear in the immediate descendants but continue to evolve although their adaptive significance does not become apparent for a long time. Anticipations and aristogenes can be considered as a case of parallelisms and/or convergences, which have always attracted orthogeneticists.

Keywords Anticipation · Aristogenes · Parallelisms · Convergences · Adaptation

Critics of Darwinism often noted that the process of adaptation concerned the modification of existing organs. New organs or evolutionary novelties appear to arise spontaneously. At the same time, the phenomenon of anticipation of characters or phylogenetic acceleration is known: organs or their rudiments evolve at an early stage of a group's evolution, disappear at the next stages and later re-emerge either as adaptations or as a mark of a new evolutionary stage. Initially, the term "anticipation" was used in a narrower sense, to denote the features of young organisms disappearing with age but appearing in distant descendants. This phenomenon was first shown in the shells of ammonites and other cephalopods. Later it was revealed in the evolution of ungulates: temporary teeth of horse ancestors were more complex than permanent ones and similar to the teeth that were to evolve later (Pavlov 1901 cited in Berg, 1977).

Evidence of anticipation was also found in embryology. The larvae of some species have features of a higher organization, which disappear in adults. Karl Ernst von Baer was one of the first to note this. Objecting to what was later called the biogenetic law or the law of recapitulation (ontogenesis is a reflection of phylogenesis), he wrote: "If this law were true, animals would not have passed in their development stages characteristic of higher animals. However, one may find quite a lot of just such cases. Of course, we would not find them in the development of man because there are no forms standing above him but other mammals can provide a

I. Popov, *Orthogenesis versus Darwinism*,
https://doi.org/10.1007/978-3-319-95144-7_11

number of such examples. In all of them jaws are as short at the early stages of development as they are in adult humans. The parietal ridge develops very late in all animals that have it, although it is absent in higher forms; the brain of birds during the first third part of the embryonic development is much more similar to the mammalian brain than in adult birds. Initially the eyes of chicken are located closer to each other than at a later stage, and resemble a human face. Young lizards have very large brains. The tadpole has a true beak, like a bird, and before it loses the tail, its intestine is as long as it is only in some mammals in adult condition. Centipedes, mites and water mites, when they leave eggs, have only three pairs of legs like insects transforming in adult condition" (Baer 1828 cited in Kryzhanovsky, 1939, p. 287). Berg interpreted phylogenetic acceleration or anticipation in a broader sense including cases when lower forms (young or adults) showed features of a higher organization that would not be present in their immediate descendants but would manifest themselves in organisms at a much higher evolutionary level. For example, bennettites, a groups of extinct gymnosperms, anticipated flowering plants in many respects. Their reproductive organs were outwardly similar to a large flower. These "flowers" were probably pollinated by insects. At the same time, bennettites are usually considered as a "side branch". They did not give rise to flowering plants. Certain "anticipatory" features were also found in other gymnosperms. Most gymnosperms are dioecious but a gnetaceous plant welwitschia (*Welwitschia mirabilis*) has a rudimentary barren ovule on male "flowers". Gnetales are also a "side branch", not the ancestors of flowering plants. Yet welwitschia imitates flowering plants by a useless rudiment of an organ.

A phenomenon similar to anticipation is the emergence of "aristogenes", that is, organs useless for an organism but creating preconditions for the origin of "useful" organs in its distant descendants. Unlike "anticipatory" features, they do not disappear in the immediate descendants but continue to evolve although their adaptive significance does not become apparent for a long time. George Simpson, a determined critic of orthogenesis, acknowledged that this phenomenon was a very strong argument in its favour (Simpson 1944).

Some characters of modern organisms indicate the past existence of aristogenes because the initial steps of their evolution seem to have been useless at that time. John Willis called such cases "morphological puzzles" (Willis 1940). One such puzzle are sundews (*Drosera*). Complex tentacles of these insectivorous plants have biological sense only in a fully developed condition. Rudimentary tentacles, which cannot be used for capturing insects, are apparently useless. This means that they have evolved either as saltations, without any intermediate forms, or from aristogenes present in ancestors of sundews. Insectivorous pitcher plants *Nepenthes* provide a better illustration of evolution of insect-capturing tools than *Drosera*, with various degrees of development of these features being observed. In this case, too, initial forms seem useless but exist nevertheless (Willis 1940).

Similar cases have been described in palaeontology. Surprisingly, some of them were given by Darwinists—with no comments or with some comments quite in contrast with the ideas of natural selection. One such example is the analysis of the origin of predatory teeth conducted by Davitashvili. Underdeveloped predatory teeth of the ancestors of carnivorous mammals seem a maladaptive feature. They were less useful for cutting and breaking than the front teeth. Nevertheless, these teeth continued to evolve because their owners "were groping for a new way" for their distant descendants. This "lucky chance" opened new vistas for the development of a progressive group—predators (Davitashvili 1948. S. 390). Effectively, this means that the ancestors of carnivorous mammals did not become extinct ceding place to animals with better teeth even though natural selection, as Darwinists claim, is a very delicate mechanism and even a minor selective advantage ensures the success of a new character. In other words, an apparently maladaptive feature was evolving for a long time.

Anticipations and aristogenes can be considered as a special case of parallelisms and/or convergences, which have always attracted orthogeneticists. Parallelisms and convergences have been referred to as the most sophisticated evolutionary phenomena (Dubinin 1966; Tatarinov 1976). Perhaps their complexity is primarily associated with the lack of a commonly accepted definition. Parallelism is usually considered as a similarity of related groups of organisms, while convergence is a similarity of unrelated ones. However, it is often difficult to distinguish related and unrelated organisms. Intermediate cases between parallelisms and convergences have been described. George Simpson considered similarities of marsupials and placental mammals as such an intermediate case. In both these groups, there are or there have been wolves, martens, moles, jerboas, mice and even sabre-toothed tigers. On the one hand, marsupials and placental mammals are two different subclasses, whose evolutionary paths diverged a long time ago. So, their resemblance can be called a convergence. On the other hand, both are mammals. The ability to evolve in similar directions was likely to be present in their common ancestor. This means that their similarities can be explained by relationship, and this case should be then considered as a parallelism (Simpson 1944).

The difference between these two phenomena may be as follows: during evolution of convergent groups there was a stage where the similarity between them was absent while parallel groups had similarities over the entire period of their evolution. In practice, however, it is difficult, if not impossible, to reveal gaps in the evolution of similar characters, and so this criterion is rarely used. Other theoretical criteria were also proposed for the distinction between convergences and parallelisms such as whether the changes were superficial or profound, whether they involved analogous or homologous organs etc. (Ryzhikov and Oshmarin 1985; Tatarinov 1985). However, as soon as it comes to the analysis of concrete examples, the boundary between these phenomena becomes hazy.

Difficulties with the definition of convergence and parallelism are closely related with the fact that their mechanisms are difficult to explain. A traditional explanation is that convergences evolve in response to adaptation of organisms to the same conditions, while parallelisms arise from the similarity of the hereditary basis, which may be enhanced by adaptation. However, these explanations do not work in the numerous cases when similarities between organisms are due neither to adaptations nor to a close genetic relationship. For instance, in different groups of lobe-finned fish rhombic scales are replaced by rounded ones, teeth evolve according to the same pattern etc. (Vorobyeva 1977). Wing venation is similar in several unrelated taxa of insects (caddis flies, moths, treehopper), although its details are unlikely to be adaptive (Grodnitsky 1995).

To explain "inexplicable" parallelisms, the hypothesis about the importance of "non-sexual" exchange of genetic material was sometimes brought up. Went developed these ideas based on data about "parallel evolution" of plants (Went 1971). He pointed to a number of features that evolved in parallel in plant groups that were not directly related such as an "eucalyptus" type of leaves in Australia, red colour of autumn leaves in the Americas and in Japan, cushion-like plants in Patagonia, tubular flowers pollinated by insects in the South-West of USA etc. According to Went, these cases are explained by the fact that certain segments of chromosomes exist in a certain region and can be transferred to numerous plant species: by mating in closely related species and by viruses in unrelated ones. Went pointed out that some plant viruses such as the virus of tobacco mosaic can infect a wide variety of plants. According to him, the success of the selection and breeding work of Ivan Michurin and Luther Burbank was due to "a casual non-sexual transfer of genes" because they bred plants in large numbers.

For an orthogeneticist, similarities between organisms than cannot be explained by relationship or adaptation are a clear evidence of regularities, directionality and predetermination, in short, evolutionary laws. In this light, there is no difference between convergences and parallelisms. The choice of the word is a matter of taste. Berg, for instance, preferred convergences, while Cope used the term parallelisms.

Orthogeneticists discovered parallelisms inexplicable in terms of natural selection theory almost everywhere. Mimicry, when a defenceless animal imitates the one that can defend itself, is a classic example of adaptation. Darwinists often mentioned it as a case for natural selection. On the other hand, Eimer, Stresemann (in the beginning of his carrier), Berg and other authors claimed that mimicry was just a display of parallelism irrespective of natural selection. They pointed out that the imitator and the object of imitation do not even always inhabit the same area. Moreover, the benefits of imitation are uncertain. There are many stingless insects that look like wasps or bees. However, there are also predators eating wasps and bees. Imitation is meaningless: any insect can be eaten anyway.

Parallelisms of unclear origin were also found in biogeographic studies. There are cases when very similar animals inhabit distant areas separated by numerous geographic barriers. Daniele Rosa wrote about discontinuous areas of tapirs and alligators (see above). With some stretching, one can probably assume that tapirs and alligators had continuous ranges in a relatively recent past. There are, however, more complex distribution patterns. Iguanas and agamas are very similar groups of lizards. Agamas inhabit Africa and Asia, while iguanas live mainly in America. Yet some iguanas live at Madagascar and Fiji where, it would seem, agamas should live. All areas around Madagascar are inhabited by agamas. Leon Croizat explained this distribution pattern as follows: initially some "pre-agama-iguana" distributes over the entire area and then it evolves, according to certain internal regularities of variation. Since these regularities are common for the initial form, similar forms arise everywhere (Croizat 1962). Vandel described a similar case. There are two similar species of woodlice, one in Nepal and the other in the western part of Australia. They were separated geographically 135–140 million years ago but are very similar at present. Vandel concluded that neither isolation nor selection under the conditions of different habitats have changed these woodlice; only orthogenesis did (Vandel 1972 cited in Nazarov, 1984, p. 99).

Parallelisms and convergences may be difficult to interpret because they cover a range of phenomena. Yuri A. Filipchenko made a bold attempt to classify parallelisms (Filipchenko 1924). He divided them into three types: genetic, anatomical and histological. Genetic parallelisms are those where external similarity is based on similar heredity. Vavilov's law of homological series provide the best examples of genetic parallelisms: related plants vary in a similar manner. On the other hand, if parallelisms affect characteristics underlying the differentiation of organisms into orders, families, classes and phyla, these cases cannot be, according to Filipchenko, explained by a similar heredity. In such cases we deal with anatomical parallelisms. For example, numerous parallel groups of relatively high taxonomic level have been described among extinct cephalopods, in particular, nautiloids and ammonoids. There are among them forms with straight, curved, spiral shells. These features are probably determined by dozens of genes; distant groups are unlikely to have them in common. And even if this were the case, parallelisms like these are not limited to nautiloids and ammonoids. Similar series of shells can be found among foraminifera, a group of unicellular organisms entirely unrelated with cephalopods. (Actually, before the studies of Félix Dujardin [1835] foraminiferans were attributed to cephalopods. Their very name was coined by Alcide d'Orbigny (1826) by analogy with Siphonifera—nautiloids and ammonites [cited in Filipchenko 1924]). The shells of foraminifera are very diverse, much more so than those of cephalopods, but similar series from straight to spiral shells can be found among them too.

How can we explain these facts? If this is convergence, it arose in different groups as an adaptation to the same conditions. However, there are no reasons to believe in the adaptive significance of minor details of shell shape. Therefore, Filipchenko preferred another explanation: "the laws of growth and gradual involution of an initially straight shell are probably the same regardless of the shell's owner, be it a unicellular rhizopod […] or a highly organised representative of the molluscan phylum" (Filipchenko 1924, p. 253).

The third type of parallelisms, histological parallelisms, are also difficult to explain by the similarity of genes. Filipchenko referred to an extensive research of parallelisms in the structure of tissues conducted by Zavarzin. The main conclusion based on the analysis of histological parallelisms was: "The similarity of structures of similar organs is an initial property developing in all forms in parallel and independently of genetic relationships"; at the same time, "as the complexity of a structure increases, its elements are simplified" (Zavarzin cited in Filipchenko 1924, p. 255).

Interestingly, Filipchenko's analysis of parallelism is in many respects similar to that conducted by the theorists of Darwinism. So, Leonid P. Tatarinov suggested (in all probability, independently of Filipchenko) four types of parallelisms: genotypic, morphological, tissue and biochemical (Tatarinov 1976). The former three were very similar to the types suggested by Filipchenko. Tatarinov (as well as Schmalhausen) arrived at similar conclusions as Filipchenko regarding the nature of non-genetic parallelisms although, of course, he worded them in such a form that it would not resemble orthogenesis or other heresies. Tatarinov wrote about the similarity of genes: "It is unlikely that it can explain more than a similarity in single characters. Schmalhausen showed that similarity is based not on genotypic but on morphogenetic commonness, that is, on the common mechanisms, which may have a different genetic basis" (Tatarinov 1976, p. 191). At the same time he noted that "…the causes of broad occurrence of parallelisms in evolution are unclear. Given an indefinite nature of inherited variability, it would be more natural to expect new characters to be original even in the case when related organisms adapt to similar conditions of the environment" (Ibid., p. 190). However, "as a rule, the majority of close branches in all groups of animals always evolve in parallel unless these branches originated as a result of transition to an absolutely new environment" (Schmalhausen 1947, p. 490 cited in Tatarinov 1976, p. 190). These statements by Tatarinov and Schmalhausen seem to imply that just a little further research and we would understand how parallelisms evolved. Such hopes are unfounded. The mechanisms of parallelisms and convergences cannot be explained by the theory of natural selection.

Finally, it is difficult to say whether convergences (similarities) of divergences (differences) are more important in evolution. If we focus on similarities between groups of organisms, their diversity seems surprisingly and inexplicably ordered, and the arguments of orthogenesis begin to seem convincing. However, whatever explanation of similarities (convergences) we can find, differences (divergence) are not to be disregarded. According to Darwinism, evolution is mostly divergent, while according to orthogenesis it is mostly convergent. What this "mostly" means, is difficult to evaluate objectively. In studies on particular taxa arguments in favour of both viewpoints can be found. For instance, there is an article entitled *Nonconvergence in the evolution of primate life history and socio-ecology* (Kappeler and Heymann 1996). Its authors considered primates to be a convenient object for testing the "convergence hypothesis": similar abiotic conditions should have resulted in similar adaptations in different primate groups. They compared the primates of Africa, Asia, South America and Madagascar by four characters (body weight, activity pattern, diet and group size) and discovered, in their opinion, the lack of convergence where convergence should have been found. Primates in each of these regions had some features distinguishing them from the others. For instance, there are no primarily folivorous primates in the Neotropics, there are many nocturnal primates on Madagascar etc. The authors explain these differences by the past role of adaptive forces, which are currently unknown (and might never be).

A deeper study of the relative role of convergences and non-convergences in evolution was attempted by Janet and Pat Willmer (Moore and Willmer 1997). Before discussing this role, they addressed the methods used to study convergences. Having analysed modern taxonomic methods ("traditional, phenetic and cladistic"), they concluded that all of them tended to conceal convergences or did not provide enough information to determine how often they arose. Nevertheless, numerous convergences are known. Moore and Willmer found that convergent evolution took place in all major groups of invertebrates. For example, forms without shells evolved in eight orders of gastropods, and series illustrating the gradual loss of shells can be constructed. In this case convergences seem to be obvious. However, studies of convergences often come to a dead end: to identify them, one has to establish phylogenetic relationships, and to establish phylogenetic relationships, one has to identify convergences. It is difficult to determine whether similarity results from a relatively recent common ancestor or from independent evolution. For example, the phylum of coelenterates is composed by swimming and sedentary forms (jellyfish and polyps) but it is unknown which of them is primary. Accordingly, it is unclear how one should look for convergence among them. Moore and Willmer mainly tried to understand to what extent convergences were common. Unfortunately, this question remained virtually unanswered, the answer being that "convergent evolution is much more important than is generally recognized" (Moore and Willmer 1997, p. 3).

To sum up, it is at the moment a matter of preference whether to consider evolution as predominantly convergent or predominantly divergent. The idea of "mostly converging" evolution is appealing. Armed with it, one may set out to put order into the diversity of organisms with the hope of finding a "periodic system" of organisms. This is how the periodic law was discovered in chemistry. However, biology followed a different way indicated by Darwin: emphasising an unlimited diversity of living organisms and the possibility of origin of any organism adapted to a particular environment. **Henri Bergson** (1859–1941) noted that in the context of Darwin's theory living matter is likened to a liquid taking the shape of the container it is poured in. The same glass can be used for wine or water, and the same environment would shape different organisms to adopt a desired shape. Bergson himself expressed an opposite opinion: "The circumstances are not a mold into which life is inserted and whose form life adopts […] There is no form yet, and the life must create a form for itself, suited to the circumstances which are made for it." (Bergson 1998, p. 58). So, in Bergson's opinion, evolution of organisms is a creative process.

Orthogeneticists were opposed to both these viewpoints. Focussing on the limitations of the organisms' "creativity", they tried to prove that there are constraints on variation and that certain "bans" determine the direction of evolution.

References

Baer KE von (1828) cited in Kryzhanovsky SG (1939) The principle of recapitulation and the conditions of historical understanding of development (an essay of the theory of historical heterogenesis) (Printsip rekapitulyatsii i usloviya istoricheskogo ponimaniya razvitiya [ocherk teorii istoricheskogo geterogenezisa]). In: In memory of academician A. N. Severtsov (Pamyati akademika A. N. Severtsova), v 1. AN SSSR, Moscow, Leningrad, p 281–383

Bergson H (1998) Creative evolution (trans: Mitchell A). Dover Publications Inc., Mineola/New York

Croizat L (1962) Space, Time, Form: The Biological Synthesis. Published by the author, Caracas
Davitashvili LS (1948) The history of evolutionar palaeontology from Darwin to our days (Istoriya evolyutsionnoy paleontologii ot Darvina do nashikh dney). AN SSSR, Moscow/Leningrad
Dubinin NP (1966) The evolution of populations and radiation (Evolyutsiya populyatsiy i radiatsiya). Atomizdat, Moscow
Filipchenko YA (1924) On parallelism in the living nature (O parallelizme v zhivoy prirode). Uspekhi sovremennoy biologii 4:242–257
Grodnitsky DL (1995) The problems of functional interpretation of some identical morphological structures on insect wings and an explanation of secondary similarity of organisms (Problemy funktsional'noy interpretatsii nekotorykh odinakovykh morfologicheskikh struktur na kryl'yakh nasekomykh i ob'yasnenie vtorichnogo skhodstva organizmov). Zhurnal obshchey biologii 56(4):438–449
Kappeler PM, Heymann EW (1996) Nonconvergence in the evolution of primate life history and socio-ecology. Biol J Linn Soc 59(3):297–326
Moore J, Willmer P (1997) Convergent evolution in invertebrates. Biol Rev Camb Philos Soc 72(1):1–61
Pavlov AP (1901) The lower cretaceous of Russia and its fauna (Le Crétace inférieur de la Russie et sa faune). Nouv Mém Soc Nat Moscou 16(3): 1–87 cited in Berg LS (1977) Works on the theory of evolution (Trudy po teorii evolyutsii). Nauka, Moscow
Ryzhikov KM, Oshmarin PG (1985) Convergencies in helminths and the question of relationships between the notions “convergence” and “parallelism” (Konvergentsii u gel'mintov i vopros o vzaimootnoshenii ponyatiy «konvergentsiya» i «parallelizm»). In: Morphological studies of animals (Morfologicheskie issledovaniya zhivotnykh). AN SSSR, Moscow, pp 179–189
Schmalhausen, 1947 cited in Tatarinov LP (1976) Morphological evolution of teriodonts and general questions of phylogenetics (Morfologicheskaya evolyutsiya teriodontov i obshchie voprosy filogenetiki). Nauka, Moscow
Simpson GG (1944) Tempo and mode in evolution. Columbia University Press, New York
Tatarinov LP (1976) Morphological evolution of teriodonts and general questions of phylogenetics (Morfologicheskaya evolyutsiya teriodontov i obshchie voprosy filogenetiki). Nauka, Moscow
Tatarinov LP (1985) Palaeontology and the theory of evolution (Paleontologiya i teoriya evolyutsii). In: Morphological studies of animals (Morfologicheskie issledovaniya zhivotnykh). AN SSSR, Moscow, pp 229–427
Vandel A (1972) The distribution of oniscoides (terrestrial isopod crustaceans) and continental drift (La répartition des Oniscoïdes [Crustacés Isopodes terrestres] et la dérive des continents) Comptes rendus de l'Académie des Sciences 275: 2069–2072 cited in Nazarov VI (1984) Finalism in the modern evolutionary teaching (Finalizm v sovremennom evolyutsionnom uchenii). Nauka, Moscow
Vorobyeva EI (1977) Morphology and characteristics of evolution of crossopterygian fishes (Morfologiya i osobennosti evolyutsii kisteperykh ryb). Trudy PIN, 163. Nauka, Moscow
Went FW (1971) Parallel evolution. Taxon 20(2/3):197–226
Willis JC (1940) The course of evolution by differentiation or divergent mutation rather than by selection. Cambridge Univ press, Cambridge

Chapter 12
Constraints on Variation

Abstract The actual number of variations is smaller than the expected one. There are no blue-eyed fruit flies, no viviparous birds or turtles, no hexapod mammals etc. Such observations provoke non-Darwinian evolutionary concepts. Darwin addressed the contradictions between his model of random variability and the existence of constraints but could not explain them convincingly. Modern Darwinism cannot not solve this problem, either, and the question persists.

Keywords Constraints · Variation · Senescence · Snouters · "Non-ageing" animals

Darwin believed that selection had extremely broad possibilities because variability always provided enough material for it. This viewpoint is still popular. Several years ago the media reported that a new race of carps with fur instead of scales was bred. Some people believed in these furry carps.

However, Darwin also described cases when breeders did not succeed in getting the desired variations. He wrote that Sir George Seabright performed long-term crossing experiments in order to obtain intermediate races between two strikingly different species (he did not mention what species they were) but these experiments failed (Darwin 1859). Now that Mendel's laws are known, this does not seem surprising. It is clear that the crossing of two pure lines results in a heterozygote; when the heterozygotes are crossed, they split and will never form a new pure line. The mechanism is known but the fact remains: it is impossible to create a race with the desired properties in this case. On the one hand, breeders seem to inspire life into a desired image, on the other hand, they merely sort out the material supplied by nature. There is nothing to be done if the desired material is unavailable. One has to be content with what one has and pretend that that was what one had wanted all along. It is due to this logic that some people considered Darwin's theory as an illusionist's trick. A Russian philosopher **Nikolay Ya. Danilevsky** (1822–1885) in his *Critical Research of Darwinism* (1885) likened Darwin's theory to the proof of the possibility of existence of two right angles in a triangle. Such a proof was presented by some mathematicians as a joke, and it was difficult to find a catch in it.

Danilevsky denounced Darwinism in all its aspects. His criticism was not always consistent, and there was little he could offer in return (though this impression may

I. Popov, *Orthogenesis versus Darwinism*,
https://doi.org/10.1007/978-3-319-95144-7_12

be partly due to the fact that Danilevsky's work was left unfinished after his sudden death). Nevertheless, Danilevsky found some interesting examples, which certainly deserve mention in a discussion of non-Darwinian concepts. His interpretation of variability of domestic birds is especially enlightening. Danilevsky was quite indignant about what he considered as Darwin's mistakes in the analysis of geese, hens and pigeons. Darwin wrote that hens were highly variable because man had been selecting them for poultry, eggs, decoration or cock-fighting. Pigeons were also variable because they had been bred as decorative birds. But geese were not—according to Darwin, because they had only been bred for their meat. Danilevsky was outraged: "So, were geese only to fall into the hands of whimsical enthusiasts, we would have had geese with peacock tails, geese with coloured feathers, crested geese and dwarf bantam-like geese. If geese had fallen into the hands of bird-fighting enthusiasts, would they have acquired spurs or some other weapons?" Danilevsky claimed that "Darwin confuses the cause and the consequence: the goose has remained unchanged not because it was not chosen as an decorative bird valued for the beauty and strangeness of shape and plumage; on the contrary, it did not become decorative bird like pigeons and hens because it had always been non-variable by nature!" (Danilevsky 1885, p. 203). Danilevsky added that in some parts of Russia goose-fighting was popular alongside with cock-fighting but geese did not acquire either spurs or any other weapon. (This case was apparently unknown to Darwin.) Danilevsky also pointed out that pheasants and peacocks remained unchanged though they had been bred as decorative birds and, according to Darwin, should be expected to be variable (Danilevsky 1885).

In fact, Darwin did note that geese had "a non-variable nature". He also gave some other examples of relatively low variability. Nevertheless, he always tried to explain them in terms of natural selection. For instance, he noted that cats and donkeys were less variable than dogs and horses, respectively, but immediately offered a selection-related explanation. Cats wander at nights and in general behave as they please, so it is difficult to control their crossings. Donkeys are mainly used on poor farms, which struggle to survive and cannot properly engage in breeding. So, Danilevsky's indignation on the matter was not entirely unfounded.

The authors of the evolutionary synthesis generally agreed with Darwin in the matter of constraints on variation claiming that variability provided an almost unlimited material for selection and that the differences in variability were mainly due to the influence of the environment. So, Ernst Mayr (1942) noted that differences in the linear size of individuals from different populations made up less than 8% in birds, and several hundred per cent in terrestrial invertebrates. He concluded that the variability of the latter was higher than that of birds and that this was so because the former were more subject to environmental influence. Structural differences between birds and invertebrates were disregarded. At the same time, he implied that abundant material for the action of the environment can be found in both these groups.

However, attempts to validate this approach often come to a dead end. Let us consider an example from Dobzhansky (1937). Twelve races of the bird called the golden whistler (*Pachycephala pectoralis*) were described on Pacific islands. To characterize them, Dobzhansky composed a table with combinations of five characters, each of which could have two variations (Dobzhansky 1937, p. 54). Eight such combinations were found in nature. Mayr (1942), in order to demonstrate a high variability of birds, cited this example as a case in point noting that almost all possible combinations were found. Actually, the number of possible combinations is 32, and eight is certainly not "almost 32". However, the belief in an unlimited variation had already been so firmly established that the absence of 24 variations passed unnoticed. Nobody asked whether the non-existing variants were eliminated by natural selection or whether they had never existed because variability was limited (Kovalenko and Popov 1997).

This example was removed from later works by Dobzhansky (except for the second edition of *Genetics and the Origin of Species* in 1941) and Mayr. In fact, discussions of the degree of variability tend to disappear altogether—apparently, they became inconvenient. It was suggested that "it is less important for the understanding of evolution to know how genetic variation is manufactured than to know how natural selection deals with it" (Mayr 1963, p. 25). Darwin was of the same opinion: "variability sinks to a quite subordinate position in importance in comparison with selection" (Darwin 1883, p. 236). Satisfied by these ideas, the architects of the evolutionary synthesis and their followers chose to ignore the issue of constraints. Similarly to Darwin, they considered this topic as insignificant. In their large treatises the discussion of constraints on variation occupies 1–2 pages at best (Futuyma 1998).

Curiously, several Darwinists did address the problem of constraints and collected examples of them. **Nikolay N. Vorontsov**, an vocal supporter of evolutionary synthesis in Russia, liked to speculate about evolutionary bans. For instance, he indicated that cartilaginous fishes are "prohibited" from inhabiting freshwater basins while amphibians are "prohibited" from inhabiting seas—much to the merriment of some opponents. Eduard Akhnazarov noted that Darwinists analysing evolutionary constraints score an own goal. The situation was all the more amusing as the examples of "prohibitions" collected by Vorontsov were not the most felicitous ones. Another Russian Darwinist, Leonid P. Tatarinov, pointed out that some cartilaginous fishes do live in freshwater, while some frogs tolerate considerable salinity. Vorontsov could actually use these instances of evolution overcoming constraints to support his Darwinian views. However, Vorontsov, who was intolerant of criticism, persistently stressed his viewpoint, maintaining that exceptions found by Tatarinov were insignificant. Actually, Tatarinov himself also scored an own goal, as he was interested in parallelisms, another source of orthogenetic heresies.

These examples show that ideological opponents can find much in common once they stop arguing in terms of paradigms or theories and start analysing concrete phenomena. Both orthogeneticists and Darwinists have described constraints on variation. In the simplest case they concern the development of a character. For example, even the strongest selection pressure for an increase in the size of on bean seeds would not result in apple-sized beans. As shown in famous experiments of Wilhelm Johannsen, selection for several generations yields pure lines after which further selection becomes fruitless (Johannsen 1903).

The absence of certain kinds of variation also highlights the limits of selection possibilities. As often as not, plant breeders fail to completely eliminate unwanted substances from a plant. For example, one can reduce the levels of alkaloids in lupine seeds or gossiol in cotton seeds but not remove these substances altogether (Goldovsky 1973). Apparently, there is no lupine without alkaloids and no cotton without gossiol. In the same way, one cannot get fruit flies with blue or green eyes (Timofeeff-Ressovsky et al. 1977).

Darwinists were familiar with these or similar examples but generally left them with little comment. Like Darwin himself, they preferred to deal with more understandable topics. In an address delivered before the International Congress of Zoologists in 1895, August Weismann presented an elaborate metaphor comparing possible variations to railroad tracks, impossible variations, to stretches without tracks and the locomotive travelling on the tracks, to the cause of variation. The destination of the train is determined not by the tracks and not by the engine but by the engine-driver—usefulness in the theory of selection (Weismann 1896). These speculations, however picturesque, seem like an attempt to get rid of uncomfortable questions about the constraints on variation. Besides, though a metaphor, of course, cannot be used as an argument, one cannot help noticing that this one can be successfully reverted to illustrate the opposite opinion: indeed, the role of the driver in determining the movement of the train is quite small.

Another apostle of Darwinism, Thomas Huxley, deliberately reduced such cases to absurdities. "A whale does not tend to vary in the direction of producing feathers, nor a bird in the direction of developing whalebone" (Huxley 1894, p. 181). In a similar vein, Nikolay Timofeeff-Ressovsky remarked that man was not predisposed to vary in the direction of variation of the tail brush (Timofeeff-Ressovsky et al. 1977). Witticisms aside, it is evident that there are no such tendencies because whales have no feathers and man has no tail. On the other hand, fruit flies have eyes, which tend to vary in colour, but no blue-eyed or green-eyed fruit flies have ever been seen.

In addition to these constraints or the "lack of developmental prerequisites", there are also "prohibited" combinations of characters. Georges Cuvier famously joked that devils devouring sinners do not exist because horned and hooved creatures can be only herbivorous. To note, while speculations on modules and modularity have recently become popular in biology, impossible combinations of modules do not generally attract much interest although they apparently exist. No selection pressure, however strong, will bring to life creatures in paintings and drawings by Hieronymus Bosch (Fig. 12.1). Impossible combinations of features have also been described in existing organisms. For example, there are the so-called tragant life forms among the legumes, when a plant looks like a pillow with thorns. For some reason, this life form is incompatible with an annual growth pattern. For instance, among plants of the genus *Astragalus* there are many annual forms and many tragant forms but there are no annual tragant ones (Popov 1929).

Whenever Darwinists acknowledged the existence of constraints, they usually asserted that variability in general was still very large. A plethora of facts indicating variability served to impress the reader and stifle any suspicion that constraints

Fig. 12.1 A drawing by Hieronymus Bosch (from: Riedl 1975)

might exist. George Stebbins, apparently trying to downplay the importance of Wilhelm Johannsen's experiments on the selection of pure lines of beans, told his students that Johannsen, a Dane working in a nice and neat Danish garden, simply could not conceive of the variation possibilities of a tropical rain forest or the waste lots of subtropics (Stebbins 1980). This absurd reproach, of course, cannot be regarded as a serious argument. However variable the manifestations of life might be, they do not in itself prove that variability is infinite.

A more serious argument of modern Darwinism is the claim that a species accumulates a very large number of recessive mutations absorbing them like a sponge. They usually remain hidden but can manifest themselves at any moment under certain selection pressure (Chetverikov 1926 cited in Chetverikov, 1983). Again, however, a "very large" number is not unlimited. Neither of the arguments affects the value of the Johannsen's data or discredits the information about the absence of certain kinds of variability.

Let us assume that we have realised the scarcity of material for selection and are ready to proclaim, following Berg and his allies, that "Selection has nothing to choose from!". Yet even such an admission, according to the supporters of the natural selection theory, would not undermine its basis. "No, there is always something", they would answer, "Even if there are few variants of variability, even if there are just two, we can still choose." (Ivantsov 1923). This means that natural selection exists, and is important in evolution. Such an interpretation, though, is more similar to the one offered by orthogeneticists: selection is not a creative force but a destroying one. Indeed, an elimination of some evolutionary trends by natural selection does not disagree with the idea of directed evolution at all.

Darwinists also claim is that even if the issue of constraints is unresolved, other arguments in favour of Darwinism are so weighty that the objections of the supporters of orthogenesis can be easily dismissed. However, historical experience suggests that such an argument is ineffective. In fact, it is an instance of sweeping dust under the carpet.

Though the analysis of constraints on variation does not fit the widespread ideas about evolution, some indications of interest in this phenomenon can be found in modern studies. Research of developmental constraints is becoming quite popular even if it usually remains confined to some particular problems of morphology and embryology. For instance, the development of the head was described in an analysis of how the rudiments of its parts mutually constrain the development of each other (Kuratani 2004).

So, the analysis of constraints on variation slips into modern science through the back-door. However, it usually reverts to a Darwinian view. Many authors do not see any necessity for an addition to the standard Darwinian models. Why should we introduce new mechanisms if the old ones explain everything? The Darwinian approach has often been accused of tautology: the surviving organisms are the fittest ones, considered the fittest because they have survived. A similar tautology can be found in speculations on constraints: "whatever has not changed must have been constrained, and whatever has changed must not have been constrained" (Eberhard 2001, p. 206). For example, the males of sepsid flies possess specific moveable abdominal lobes. These structures are extremely rare in other flies. Maybe some obstacles prevent insects from producing such structures, and only a few of them could overcome them. Thus, constraints determined evolution (Wagner and Müller 2002). According to another viewpoint, these lobes evolved in sepsids under a special pressure of sexual selection which is absent in other species. This viewpoint was supported by the data on multiple (double) origin of these lobes in sepsids. Male sepsids are slowly "wandering across the phenetic landscape" under the action of selection (Eberhard 2002, p. 7).

An adaptationistic Panglossian paradigm still satisfies many authors dealing with constraints. Their logic is as follows: if there was a change, it was useful and appeared in the process of mutations and selection. For example, the trilobites' ability to enrol increased in the course of evolution, and this means a better protection against predators (Hughes et al. 1999).

Nevertheless, a promising tendency to some acceptance of criticism and some attempts to introduce reforms is nascent. At least, constraints are now considered as a possible explanation of evolutionary phenomena instead of being ignored outright. For example, the origin of phyla or body plans (*Bauplanen*) has long troubled evolutionists. The main body plans appeared hundreds of millions years ago and have been only changing after that, with no new plans emerging. Can it be that constraints are involved? Some authors believe that the idea is gratuitous. In the process of saturation and structuration of the ecological space, conditions became unsuitable for the emergence of new body plans, that is why they have not emerged. Such an interpretation at least spells some readiness to explain this phenomenon in terms of evolutionary constraints (Ciampaglio 2002, 2004).

In some cases, one admits that the lack of the material for variability was a significant factor influencing evolutionary events. For instance, the number of cotyledons varies little in plants. This has been explained both by selection and the lack of variability (Conner and Agrawal 2005). Other authors claim that both selection and the "specifics of design" should be taken into account in evolutionary explanations. Take the selection towards miniaturization of salamanders, which resulted in some significant changes in their shape. The small size of the buds of extremities led to the decrease of the number of fingers, the necessity to pack the cells in small spaces led to the changes in shape of the head and the nervous system etc. Therefore, to explain the evolution of salamanders both design limitations and functionalist interpretations should be considered (Wake 1991).

Finally, in the discussions of these two possibilities, selection and constraints, the latter have been sometimes considered as more significant. For instance, the differences in pigmentation in different segments of *Drosophila* have a strong interaction with developmental genes. These interactions are less likely to be a consequence of natural selection and should be interpreted as evolutionary constraints (Gilbert et al. 2000).

Discussions like these usually concern isolated evolutionary cases. Attempts to draw more general conclusions usually result in compromises: panselectionism is certainly is not the best position but this does not mean that it will necessary turn into panconstraintism (Arthur 2001); constraints and selection are not opposite forces but complementary ones (Fusco 2001); phenotype is a compromise between constraints and selection (Gilbert 2000); constraints could even contribute to adaptation because unconstrained variation slows down evolution (Wagner 1988).

The best way to "solve" the issue of constraints on variation is to place them in the long list of every other possible constraint: "ontoecogenophyloconstraints" (Antonovics and van Tienderen 1991); structural, functional, genetic, developmental, cellular, metabolic, maternal constraints (Hall 1992) etc. This list may be very long, as any change in one direction means a constraint in some other direction. Developing this idea, one may conclude that biology in general is a series of constraints. It is evident that in such context any kind of constraints may be considered as more important than the other. The possibility of evolution along the rails is obfuscated again.

However, these lists do contain constraints corresponding to orthogenesis. For example, Janson (1992) identified three types of constraints: 1. Phylogenetic inertia. Adaptive radiation is limited by ancestors in several ways. Closely related organisms have less time to acquire differences than distant taxa. 2. Reduced viability of transitional forms between the two adaptive optima. 3. Structural constraints (physiological, embryological or architectural features limit the scope of possible changes). In a study on convergences mentioned earlier in this book, Moore and Willmer (1997) found five types of constraints: 1. Phylogenetic or evolutionary constraints determined by an ancestral heritage and preventing essentially new evolutionary ways. 2. Development constraints, that is, difficulties in the formation of certain phenotypes due to epigenetic processes occurring during embryonic development. 3. Genetic constraints: factors limiting genetic variability (overlapping with the previous type). 4. Fabricational constraints caused by an inability to synthesise certain substances. 5. Functional constraints imposed by everything necessary for the emergence of viable forms over the entire evolution of a group.

It is easy to notice that type 3 in Janson (1992) and types 2 and 3 in Moore and Willmer (1997) are exactly the constraints underlying theoretical speculations of orthogeneticists. Both these papers are written strictly within the framework of the theory of natural selection, and neither mentions the idea of directed evolution. Yet some support of orthogenesis can be found in both.

While constraints on variation are an argument in favour of orthogenesis, information about them is confusing. Orthogeneticists drew attention to them often enough but tried to develop methods to study them and to explain their mechanisms much less frequently. An old work by **D'Arcy Wentworth Thompson** (1860–1948) on the growth and form is still of interest in this respect (Thompson 1917 [1961]). This scientist focused on the correlation between organic phenomena and mathematical patterns and physical laws describing in detail all possible limitations imposed on organic forms by their physical conditions. He examined these constraints starting from evident cases such as limitation on the size of flying animals and finishing with the minute details of the action of surface tension or crystallisation in cells and tissues. Every time it turned out that growth and form were largely determined by purely physical and chemical factors. For example, an asymmetry of the dolphin's skull corresponds to fluid flows inside and outside the body originating during its movement in water. Unlike the movement of fish, that of whales resembles the motion of the propeller, and a "helical component of locomotion" causes asymmetric flows. (Thompson mainly analysed the narwhal horn, which is formed on the right side of the head, as well as the data of Shuleikin [1935] on the locomotion of dolphins).

Spicules of sponges and radiolarians are another notable example. Polemicizing with Ernst Haeckel (1834–1919), Thompson claimed that these structures were crystals having a certain shape not because it contributed to the success in competition (as Haeckel thought) but due to their chemical and physical characteristics. Spicules cannot take any other shape because of nature of the crystal-forming substance. Thompson also thought that various kinds of Radiolaria float equally well and that it was meaningless to explain the evolution of their form in terms of

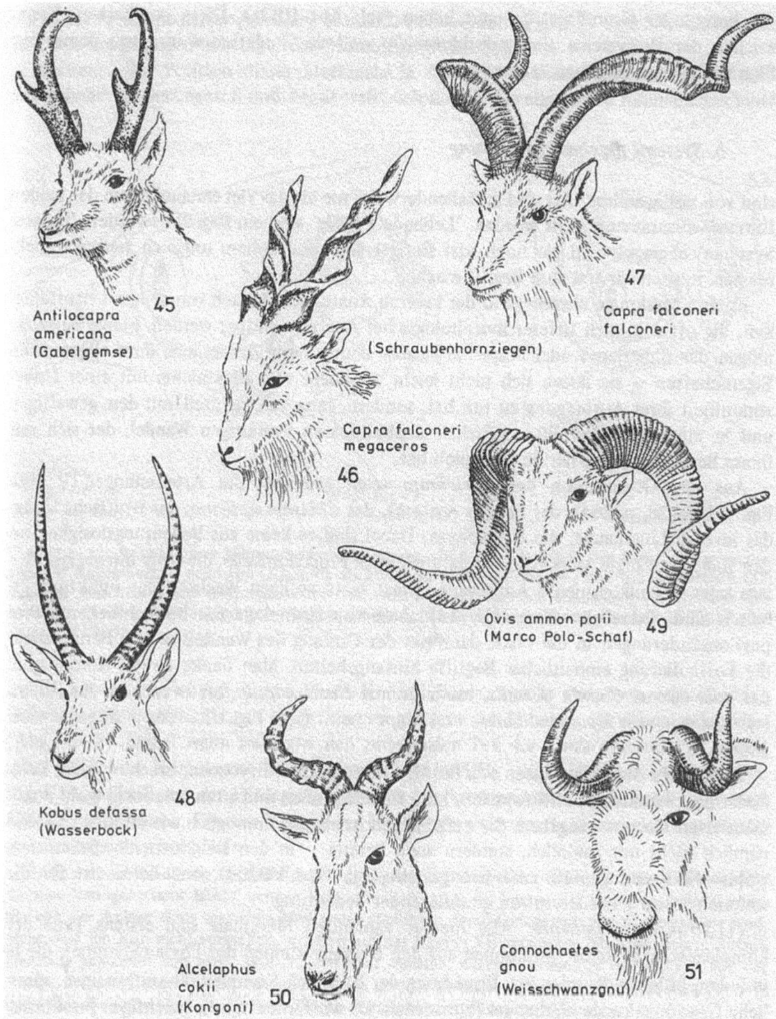

Fig. 12.2 Diversity of horns in antelopes (Riedl 1975)

Darwinian struggle for existence. So, chemical and physical constrains determine many particularities of the form of living organisms. In combination with each other, they probably form a mechanism of imposing constraints on variation resulting in directed evolution.

Modelling theoretically possible kinds of variation and comparing them with the actual ones might be promising method of analysing constraints. **Rupert Riedl** (1925–2005) in his book *Die Ordnung des Lebendigen* (*The Order of Life*) presented illustrations of horns of antelopes alongside with drawings of antelopes with imagined horns: horns thickened at the ends, horns growing downwards etc. (Riedl 1975) (Figs. 12.2, 12.3, 12.4). Indeed, estimations of the ratio of actual, rare and

Fig. 12.3 Anomalous variants of horn structure in antelopes (Riedl 1975)

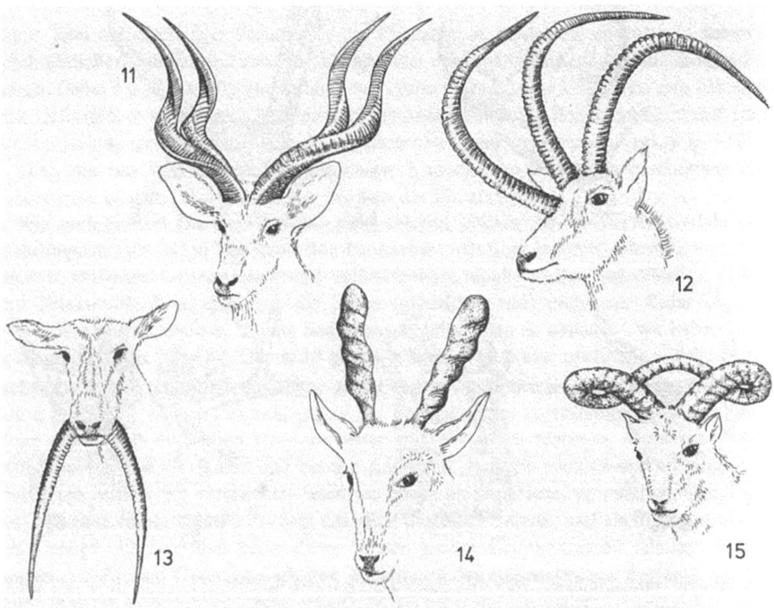

Fig. 12.4 Non-existent variants of horns (Riedl 1975)

impossible variants might elucidate the laws of variation. Such as approach was used in some morphological studies. Elena Kovalenko, who studied amphibians, elaborated the method of variability spectra (Kovalenko 1996a, b). The method involves the determination of a theoretical spectrum of variability and then of the actual ones. It was proposed for the study of variation properties: its degree, its constraints, frequency of occurrence of different kinds of variation, comparison of various species in respect of degree of variation etc. The sacrum of frogs and other tailless amphibians served as a model object, and only 45 of 288 possible variants were detected during long-term studies.

Similar results were obtained in theoretical morphology, a research trend developed by George McGhee and his colleagues (McGhee 1999). In essence, it implies the identification of a range of possible options for variation on the "adaptive landscape" and then filling it with actual options. It turns out that the actual variants fill a small part of the theoretically possible variation space.

Studies on variation spectra or morphospaces have not struck root. After the presentation of the method of variability spectra, only one other researcher decided to perform a similar study on another object (insects) (Zakharova 2002). Generally, constraints on variation remain unexplored. The researchers usually address only the actual variability, in other words, pay attention to the first of Riedl's drawings (Fig. 12.2) ignoring anomalous and non-existent variants (Figs. 12.3, 12.4).

A commendable example of an entirely opposite approach is the book *Bau und Leben der Rhinogradentia* by Harald Stümpke (Stümpke 1957 cited in Stümpke, 1993; published in English as *The Snouters: Form and Life of the Rhinogrades*). It describes a new order of mammals, the Rhinogradentia, endemic in remote Pacific islands of Hy-Yi-Yi. Their distinguishing feature is the nasarium, a structure derived from the nose. While primitive rhinogrades have relatively normal noses, the nasaria of advanced representatives have acquired a vast array of incredible shapes associated with performing the functions of motion, catching insects, fishing, digging etc. (Fig. 12.5). Sadly, Hy-Yi-Yi islands sank into the ocean during secret tests of atomic explosions nearby together with all their inhabitants as well as the collections of the Darwin Institute of Hy-Yi-Yi.

The fictitious Rhinogradentia were the brainchild of a German zoologist **Gerolf Steiner** (1908–2009), who wrote under an alias of Harald Stümpke. His book was published by Gustav Fischer Verlag, a venerable German publishing house, and had all the features of a scientific publication, complete with descriptions of families, genera and species, phylogenetic schemes and maps. References to imaginary studies were mixed with those to works by Bernard Rensch and Adolf Remane. In the years following the book's publication even some biologists apparently believed that the rhinogrades had really existed. It is difficult to say how many readers were actually deceived but the fictitious snouters have become quite popular. The book was translated into several languages, and the joke was rebranded several times: Soviet zoologists "found" relatives of the Rhinogradentia close to the Antarctic, stuffed rhinogrades were on display in various museums, a meeting with a "Professor Stümpke" was organised in Germany etc.

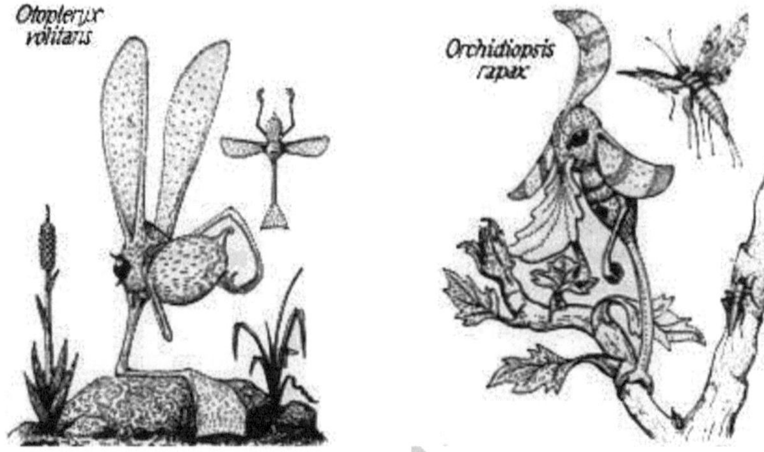

Fig. 12.5 The snouters: representatives of the order Rhinogradentia (Stümpke 1957)

The success of the Rhinogradentia hoax emphasises the prominent place occupied by the idea of an unlimited variability in the common scientific mind. According to Darwin's theory, there are no reasons why snouters should not have evolved provided a selection pressure for unusual noses was strong enough.

The dominance of the idea about an unlimited variability has important consequences for activities related to elimination of ageing, much in the same way as the prevailing ideas on adaptation do (see the discussion in section 7.3). Something like a quest for the fountain of eternal youth is blooming in modern medicine. Nobody seems to doubt that it is possible not only to cure senile diseases and improve the health of senior citizens but to cancel the very program of ageing. It would seem that there were no evolutionary restrictions on the way and that the achievement of this aim were merely a matter of time and money. The documents of the Strategies for Engineered Negligible Senescence Foundation (SENS) and other organizations contain an assurance that in the coming decades human ageing will be almost eliminated (http://www.sens.org). A good financial backing is the only thing needed to achieve this.

Statements like these contradict common sense which shows that everything in nature changes, ages and ceases to exist. Not only animate but also inanimate objects undergo changes: mountains crumble, waterfalls become lower, rivers slow down, lakes become shallow and turn into swamps etc. Living organisms change much faster, and manifestations of their aging are more visible. Therefore, it is difficult to believe that this process may be cancelled. The search for an eternal youth seems to belong to the past of humankind. Yet now it is being conducted with all paraphernalia of the scientific approach. At the same time, outside the academic community a variety of highly dubious "anti-ageing" products supposed to eliminate ageing are being peddled, and the fact that the possibility of elimination of ageing is mentioned in scientific studies encourages this activity.

Examples of "non-ageing" or "negligibly ageing" animals are used to support the possibility of cancelling senescence. *Some Animals Age, Some Do Not*—this phrase, which Leonard Hayflick chose as the title of a chapter in his famous book *How and Why We Age* (Hayflick 1994), has risen to the status of a dogma. It is widely accepted that some animals never die of ageing; if they die naturally, they die for different reasons. Let us examine one such claim.

"The pearl mussel, which grows throughout its lifetime, perishes because the weight of the shell valve becomes too large for the muscle keeping the organism vertically at the river bottom. The muscle also grows with age but not fast enough to keep up with the shell growth. Giant turtles, which also grow throughout their lifetime and finally lose mobility, may die for a similar reason (disproportionate growth of the shell and the limb muscles). Both these species live for more than two centuries without showing any senile features. Bowhead whale can compete with them in respect of the lifespan. The reason of its death is unknown but we know that at a time scale of two centuries spontaneous L-D isomerisation of amino acids in the proteins of the crystalline lens must exceed 20%, which would be critical for the sight of an animal. It is known that crystallins are not replaced in an animal's lifetime. This means that isomerisation of amino acids may cause the disruption of the protein structure and the loss of transparency of the crystalline lens" (Skulachev 2009, p. 128).

However, this description of the circumstances of the death of pearl mussels is far from reality. Pearl mussels die at different ages, as all shellfish do, and long-livers are rare. Their mortality is similar to that of humans: the usual lifespan is 30–50 years, a few individuals live up to 60–70 years, occasional long-livers reach the age of 100 or more (Popov and Ostrovsky 2011). The same might be true of turtles (at least, there are no data that they age and die naturally in a different way). Whales have been hunted for such a long time that the natural causes of their death are difficult to estimate. However, considering that they have a highly developed echolocation system suitable for navigation in the total darkness of the ocean depths, a decrease in visual acuity with age is unlikely to be the main cause of their death.

Similarly to the case of slow ageing of bats (Sect. 7.3), information about the so-called non-ageing animals is often unreliable. An entire list of presumably slow-aging animals has formed. It includes *Hydra*, some bivalves and some turtles, rockfish of the genus *Sebastes*, mole rats *Heterocephalus glaber* and bowhead whales *Balaena mysticetus*. Continuous growth and the capacity to reproduce during the entire lifetime as well as a long lifespan are considered as indications of the absence of ageing. These examples are so firmly established in the literature on ageing that they are now migrating from one work to another, often without references to primary sources of information. In such a case, an amplification of errors is likely. In fact, the circumstances of involvement of these animals in gerontology are sometimes questionable. For example, the list of negligibly ageing animals includes two species of American turtles, *Emydoidea blandingii* and *Chrysemys picta*. It is unclear why these two species should have been singled out of other turtles. It turns out that these "non-ageing turtles" were observed in a reserve for a considerable period. It was noted that they lived and laid eggs for a long time. Then the researcher who studied them retired, and the observations were discontinued. No data on how age affects the probability of their death are available (Finch 2009).

Rockfish made their way into the list of non-aging species in a similar manner. A manager of research projects read a popular book *How and Why We Age* by Leonard Hayflick (1994) and was attracted by an idea that some organisms do not seem to age. He turned to the problem of ageing and selected rockfish as a suitable object, long-living and easy to study (Guerin 2004). A large-scale project involving twelve American and two European universities was launched. Unsurprisingly, after these activities rockfishes were well known to gerontologists though it is quite probably that many other fish could have served as "non-ageing" research objects equally well.

One of the candidates for the list of "non-ageing" animals is the cod *Gadus morhua*. It has been harvested intensely for several centuries, and has had few chances of living up to the maximum age. However, even 100–200 years ago considerably larger individuals were caught than we can imagine now. A cod weighting 211 pounds was caught in 1895, while today 50-pound cods are regarded as large and 100-pound cods are exceedingly rare (Ellis 2004). Some sturgeons might also be considered as "non-ageing" because they may live for decades (http://genomics.senescence.info/species/) but they have also long lost the chance to reach old age, and their involvement into gerontology is problematic.

Recent reports of non-ageing fish are mostly associated with the fact that, owing to the depletion of fish stocks and the improvement of fishing techniques, many "new" fish species are being involved in commercial fishing. Previously these species were poorly known or difficult to obtain, and so had a chance to live a long life. Some rockfish are among such species because they live deep in the sea. This is also true of other fish. Orange roughy *Hoplostethus atlanticus* is an especially illustrative example. Large accumulation of this fish were found at great depths (several hundred meters) off the shores of New Zealand and Australia in the 1970s. They were exterminated very fast. Some of the individuals in these accumulations were more than 100 years old (Ellis 2004). The aging mechanisms of these fishes are regarded as something requiring special studies although there are no reasons to believe that they are fundamentally different from those of many other fish species. The only differences between "non-ageing" and "usual" fish probably concern growth and development rates and must be associated with differences in temperature and diet in their biotopes. Moreover, studies of "non-aging" fish are mainly based on tissue samples brought by fishermen. There are no data on mortality in natural environment or aquaculture.

Bowhead whale is another animal which often surfaces in discussions concerning senescence. Indeed, the age of one bowhead whale was assessed as 211 years. However, there are as good as no data on the manifestations of senescence, or absence thereof, in this individual. Its age was estimated based on the study of an eye tissue sample brought by a whaler. The only information about its ageing was the opinion of the whaler who believed this whale was old because its meat and fat were tough. There are other records of long-living whales, based on findings of fragments of old-fashioned harpoons in their tissues, but the estimates are less than

200 years. Original reports on long-living whales do not contain any conclusions about the absence of ageing in these animals (George et al. 1999).

The idea about immortality of the hydra is based on a study showing that the hydra's body has a zone of constantly dividing cells, which gradually restore worn and lost ones (Martinez 1998). The hydras were observed for four years, and their mortality remained insignificant though some age related changes were recorded nonetheless (Martinez 1998). This means there is some indirect evidence indicating that a hydra may potentially be immortal. However, its immortality has never been proven, and the author of the study admitted as much.

Age-related changes in the probability of natural death have not been studied in these cases, and so the main principle for identifying non-aging animals cannot be applied. It is substituted by others such as a long maximum lifespan, the retention of ability to reproduce and good health at an advanced age. Yet these principles are insufficient. Firstly, if the maximum lifespan of a species is longer than that of another, this does not mean that the representatives of the former can live forever. Secondly, a decreased ability to reproduce and the cessation of growth are age-related changes only in some species, e.g. humans. One cannot assess turtles and molluscs by the standards of mammals. Thirdly, sound health of some individuals of a species at an advanced age does not mean that *all* individuals do not age, all the more so as the available information is incomplete (Popov and Ostrovsky 2011).

Thus, the idea that some animals do not age is a myth that has entered modern biology and medicine. This myth is difficult to banish partly because of the prevailing Darwinian view about an unlimited variability. In this context, the issue of constraints on variation and directed evolution arises again.

References

Antonovics J, van Tienderen PH (1991) Ontoecogenophyloconstraints? The chaos of constraint terminology. Trends Ecol Evol 6:166–168

Arthur W (2001) Developmental drive: an important determinant of the direction of phenotypic evolution. Evol Dev 3(4):271–278

Chetverikov SS (1926) cited in Chetverikov SS (1983) The problems of general biology and genetics (reminiscences, articles, lectures) (Problemy obshchey biologii i genetiki [vospominaniya, stat'i, lektsii]). Nauka, Novosibirsk

Ciampaglio CN (2002) Determining the role that ecological and developmental constraints play in controlling disparity: examples from the crinoids and blastozoan fossil record. Evol Dev 4(3):170–188

Ciampaglio CN (2004) Measuring changes in articulate brachiopod morphology before and after the Permian mass extinction event: do developmental constraint limit morphological innovation. Evol Dev 6(4):260–274

Conner JK, Agrawal AA (2005) Mechanisms of constraints: the contributions of selection and genetic variance to the maintenance of cotyledon number in wild radish. J Evol Biol 18(1):238–242

Danilevsky NYa (1885) Darwinism. A critical study (Darvinizm. Kriticheskoe issledovanie). St Petersburg

Darwin Ch (1859 [1987]) The Origin of Species. (Reprint of 1st edition 1859) Suffolk: Penguin books
Darwin Charles (1883) The Variation of Animals and Plants Under Domestication. 2nd, Revised. D. Appleton, New York
Dobzhansky T (1937) Genetics and the origin of species, New York. Columbia University Press
Eberhard WG (2001) Multiple origins of a major novelty: moveable abdominal lobes in male sepsid flies (Diptera: Sepsidae), and the question of developmental constraints. Evol Dev 3(3):206–223
Eberhard WG (2002) Restraint with constraints: a reply to Wagner and Müller. Evol Dev 4(1):7–8
Ellis R (2004) The empty ocean. Island Press, Washington
Finch CE (2009) Update on slow aging and negligible senescence — a mini-review. Gerontology 55:307–313
Fusco G (2001) How many processes are responsible for phenotypic evolution? Evol Dev 3(49):279–286
Futuyma DJ (1998) Evolutionary biology, 3rd edn. Sinauer Associates, Sunderland
George JC, Bada J, Zeh J, Scott L, Brown S, O'Hara T, Suydam R (1999) Age and growth estimates of bowhead whales (Balaena mysticetus) via aspartic acid racemization. Canad J Zool 77(4):571–580
Gilbert P, Moreteau B, David JR (2000) Developmental constraints on an adaptive plasticity: reaction norms of pigmentation in adult segments of Drosophila melanogaster. Evol Dev 2(5):249–260
Goldovsky AM (1973) Biochemistry and problems of evolution (Biokhimiya i problemy evolyutsii). In: Vorontsov NN (ed) Problemy evolyutsii (Problems of evolution), 3, AN SSSR, Novosibirsk, pp 57–83
Guerin JC (2004) Ageless animals news. Inaugural newsletter. September
Hall BK (1992) Evolutionary developmental biology. Chapman and Hall, London
Hayflick L (1994) How and why we age. Ballantine books, New York
Hughes NC, Chapman RE, Adrain JM (1999) The stability of thoracic segmentation in trilobites: a case stady in developmental and ecological constraints. Evol Dev 1(1):24–35
Huxley T (1894) Darwiniana. Collected Essays. II. MacMillan and Co, London
Ivantsov NA (1923) The factors of evolution (Faktory evolyutsii). Gosudarstvennoe izdatel'stvo, Moscow-Petrograd
Janson CH (1992) Measuring evolutionary constraints: a Markov model for phylogenetic transitions among seed dispersal syndroms. Evolution 46(1):136–158
Johannsen W (1903) Heredity in populations and in pure lines (Erblichkeit in Populationen und in reinen Linien). Gustav Fischer, Jena
Kovalenko EE (1996a) An analysis of variability of the sacrum in Anura. 1. The method of analysis of variability of the sacrum in Anura (Analiz izmenchivosti kresttsa Anura. 1. Metod analiza izmenchivosti kresttsa beskhvostykh amfibiy). Zoologicheskiy zhurnal 75(1):52–66
Kovalenko EE (1996b) An analysis of variability of the sacrum in Anura. 2. Variability of the sacrum in representatives of the genus Rana (Analiz izmenchivosti kresttsa Anura. 2. Izmenchivost' kresttsa u predstaviteley roda Rana). Zoologicheskiy zhurnal 75(2):222–236
Kovalenko EE, Popov IY (1997) A new approach to the analysys of properties of variability (Novyy podkhod k analizu svoystv izmenchivosti). Zhurnal obshchey biologii 58(1):70–83
Kuratani S (2004) Evolutionary Developmental Biology and Vertebrate Head Segmentation: A Perspective from Developmental Constraint. In: Hossfeld U, Olsson L, Braidbach O, Levit GS (eds) Evolutionary Morphology from Carl Gegenbaur to the Present. Fineday Press, Saint-Petersburg, pp 243–268
Martinez D (1998) Mortality patterns suggest lack of senescence in hydra. Exp Geront 33(3):217–225
Mayr E (1942) Systematics and the origin of species from a viewpoint of zoologist. Harvard University Press, Cambridge, MA

Mayr E (1963) Animal species and evolution. Belknap Press of Harvard University Press, Cambridge (Mass)

McGhee GR Jr (1999) Theoretical morphology, New York. Columbia University Press

Moore J, Willmer P (1997) Convergent evolution in invertebrates. Biol Rev Camb Philos Soc 72(1):1–61

Popov IYu, Ostrovsky AN (2011) Differences in the lifespan of the freshwater pearl mussel Margaritifera margaritifera as evidence for the infeasibility of negligible senescence (based on data for St. Petersburg and Leningrad oblast). Adv Gerontol 1(2):191–197

Popov MG (1929) The genus Cicer and its species (Rod Cicer i ego vidy). Trudy po prikladnoy botanike, genetike i selektsii 21(1): 1–240

Riedl R (1975) The order of living things (die Ordnung des Lebendigen). Verlag Paul Parey, Hamburg and Berlin

Skulachev VP (2009) How to cancel the programme of ageing of the organism? In: (Kak otmenit' programmu stareniya organizma?) Rossiiskiy khimicheskiy zhurnal, vol 53, pp 125–140

Stebbins GL (1980) Botany and the synthetic theory of evolution. In: Mayr E, Provine W (eds) The evolutionary synthesis: perspectives on the unification of biology. Harvard Univ press, Cambridge (Mass), London

Stümpke G (1957) cited in Stümpke G (1993) Bau und Leben der Rhinogradentia. Gustav Fischer Verlag, Jena. English edition: Steiner G (1981) the snouters: form and life of the rhinogrades. University of Chicago Press

Thompson D'AW (1917 [1961]) On Growth and Form. Cambridge Univ Press, Cambridge (6th ed)

Timofeeff-Ressovsky NV, Vorontsov NN, Yablokov AV (1977) A brief sketch of the theory of evolution (Kratkiy ocherk teorii evolyutsii). Nauka, Moscow

Wagner GP (1988) The influence of variation and developmental constraints on the rate of multivariate phenotypic evolution. J Evol Biol 5:45–66

Wagner GP, Müller GB (2002) Evolutionary innovations overcome ancestral constraints: a re-examination of character evolution in male sepsid flies (Diptera: Sepsidae). Evol Dev 4(1):1–6

Wake DB (1991) Homoplasy: the result of natural selection or evidence of design limitation? Am Nat 138:543–567

Weismann A (1896) On germinal selection: a source determines determines directed variation (Ueber Germinal-Selektion: Eine Quelle bestimmt gerichteter Variation). Fischer, Jena

Zakharova EYu (2002) An analysis of regularities of phenotypical variability of eyespots in the browns (Lepidoptera: Nymphalidae: Satyrinae) (Analiz zakonomernostey fenotipicheskoy izmenchivosti glazchatykh pyaten barkhatnits [Lepidoptera: Nymphalidae: Satyrinae]). Extended abstract of Cand. Sci. (Biol.) Dissertation. Ekaterinburg http://genomics.senescence.info/species/ http://www.sens.org

Chapter 13
Direct Impact of the Environment on Evolution and the Inheritance of Acquired Characteristics

Abstract Speculations about a direct effect of the environment on evolutionary changes are closely associated with orthogenesis because accusations of Lamarckism were have often been directed at its supporters. At the same time, the ideas of Lamarck and his followers were discredited by the idea of the "inheritance of acquired characteristics". This seems to be a firmly established fallacy. Lamarck indeed wrote about exercises and action of the environment upon evolution but never used the terms "heredity" or "inheritance".

Keywords Inheritance · Action of environment · Lamarckism · Evolution · Change

Speculations about a direct effect of the environment on evolutionary changes are closely associated with discussions concerning orthogenesis. Lamarckism, although never the main point of concepts of directed evolution, has often been mentioned in this connection. Accusations of Lamarckism were one of the points of criticism directed at the supporters of orthogenesis.

At the same time, the ideas of Lamarck and his followers were discredited by the so-called problem of the inheritance of acquired characteristics. In fact, this problem seems to be a firmly established fallacy. Lamarck indeed wrote about exercises and action of the environment upon evolution but never used the term "heredity" or "inheritance". Some of his speculations sound quite acceptable. For instance, he described a possible pattern of origin of new plant species as follows: "Suppose, for example, that the seeds of a grass or any other plant that grows normally in a damp meadow, are somehow conveyed first to the slope of a neighbouring hill where the ground although higher is still rich enough to allow the plant to maintain its existence. Suppose that then, after living there and reproducing itself many times, it reaches little by little the dry and almost barren ground of a mountain side. If the plant succeeds in living there and perpetuating itself for a number of generations, it will have become so altered that botanists who come across it will erect it into a separate species. The same thing happens in the case of animals that are forced by circumstances to change their climate, habits, and manner of life: but in their case

I. Popov, *Orthogenesis versus Darwinism*,
https://doi.org/10.1007/978-3-319-95144-7_13

more time is required to work any noticeable change than in the case of plants." (Lamarck 1809 cited in Lamarck, 1914, p. 39).

Lamarck's speculations about exercises and habits of animals may seem less successful. For example, he wrote that the ancestors of snakes acquired a habit to crawl, underwent corresponding change of habits and exercises and in time transformed into elongated creatures without legs. However, as shown above, we can find very similar speculations in modern papers on evolution: the ancestors of snake first started to adapt to crawling, then lost legs and acquired an elongated body.

Reading Lamarck's treatises, we should take into account that they were written in a scientific context of more than 200 years ago. Lamarck did not know the mechanisms of heredity and probably was not even very interested in them. He described general images of evolution stressing the duration and gradual character of evolutionary changes. "The problem of inheritance of acquired characteristics" was stated much later, and its correctness is doubtful.

In the late 19th—early twentieth century experiments on transformation of plants and animals under the direct influence of the environment were popular. Nikolai Levakovsky noticed during spring floods that bramble shoots immersed in water acquire an unusual form. He performed experiments and described changes in tissues and shape of such shoots. French researchers carried out series of similar studies. Plants were transferred from Paris to the mountains, drought-resistant plants were grown in a humid environment, water plants were grown under different conditions etc. Plants varied greatly, bringing to mind evolutionary changes (Bonnier 1907). Similar experiments were performed on animals. A Swedish physiologist **Frithiof Holmgren** (1831–1897) tried to transform doves into birds of prey by feeding them on flesh. The reports on his studies were published several times in *Uppsala Läkareförenings förhandlingar* (Olsson 2003). There were also attempts to obtain carnivorous cocks (Houssay 1907).

The results of these and other similar experiments were interpreted in different ways but in general testified in favour of Lamarckism. However, August Weismann supposedly proved that that the concept of inheritance of acquired characteristics was wrong and its alternative, Neo-Darwinism, was correct. The results of his experiments on mice were considered as a strong evidence against Lamarckism. In the October of 1887 Weismann procured 12 white mice, 7 females and 5 males, and cut off their tails. A month later these mutilated mice gave birth to the offspring with normal tails. Some of them suffered the same fate as their parents: an inquisitive researcher mutilated them too. This was to be the fate of many further generations of mice. The experiments continued for several years. With the birth of each new generation of mice, it was ascertained that parents with cut tails give birth to normal mice. Weismann was not satisfied until he mutilated in this way several hundred mice (Weismann 1892). It would seem that he expected that generation to acquire a character (the absence of the tail) and the next one only to inherit or not inherit it. In fact, while the first generation acquires a character (a cut tail), the second one also acquires one (a non-cut tail). That is why the mice of the second generation live with a tail for some time until they acquire a new character and get their tail cut. An influence of the environment indeed leads to a certain result: in one case, it is the origin of tailless mice and in another case, the origin of mice with a normal tail.

From the point of discrediting Lamarckism, these experiment have a zero value. The mice suffered in vain. The fact that such a study had to be carried out at all is quite surprising. At that time, the scientific community was well aware that the practice of cutting or injuring body parts had been common among some peoples for thousands of years. Yet children born from the affected parents were never traumatised in the same manner.

Many other animals were mutilated at the hands of the supporters of Lamarckism fighting the nascent genetics. The best known case are experiments by an Austrian zoologist **Paul Kammerer** (1880–1928), who looked for Lamarckian mechanisms of variation in different animals. When a siphon is cut off an ascidian, it regenerates a new one, which is longer than the initial one. As an experimenting Lamarckian, Kammerer expected the offspring of traumatised ascidian to possess elongated siphons. In other words, he wanted the offspring only to inherit a new character without acquiring a new one (normal siphon) under the influence of "non-cutting". Unfortunately, Kammerer's desire led to an instance of wishful thinking. He claimed that ascidians originated from injured parents had longer siphons than normal ones. He also tried to collect other similar facts to prove the point.

The results were disastrous. Kammerer's experiments on transforming midwife toads (*Alytes obstetricans*) into "common" toads were the last straw. Unlike common European toads, midwife toads do not just lay eggs in water and leave them there. Male midwife toads carry fertilized eggs on their backs. Kammerer reared midwife toads under conditions of increased temperature and reported that they stopped to take care of the offspring and started to lay eggs in water. Allegedly after studying several generations of toads, he found that starting with the third generation the males of midwife toads acquired the so-called nuptial callosities, structures on limbs serving for retaining females during mating in tailless amphibians which lay eggs in water.

The scientific community questioned the credibility of this report and demanded that a midwife toad with nuptial callosities be presented. Yet all the preparations seems to be missing. Kammerer usually relied on drawings and did not care about fixing and photographing the material. With great a difficulty he procured one specimen of a "transformed" toad. Its examination revealed that a dye similar to Indian ink was introduced into the toad's leg and that no nuptial callosities were present. Kammerer declared that someone had framed him. However, some inquisitive people noticed another aspect of his work: the number of animal generations he had studied was so large that Kammerer must have started these studies as a schoolboy. The evidence was damning. The exposure of Kammerer's experiments was published in *Nature* and other journals. Finally, Kammerer committed suicide (Gaysinovich 1988; Gliboff 2006). This tragedy, stemming from a preconceived approach a scientific problem, was further exacerbated by a muddle created by the Soviet propaganda. Kammerer's experiments, which were in accord with the ideas of Lysenko, were discussed in a favourable light in the USSR. Kammerer was invited to the Soviet Union and offered every opportunity for further research. After his death, his story was presented as an instance of persecution of a great scientist by bourgeois reactionaries.

A less dramatic repetition of a similar story can be seen in vain attempts of an Australian immunologist **Edward J. Steele** to prove that Lamarck had been right. According to Steele, multicellular organisms with sexual reproduction may possess some forms of a "hidden Lamarckian inheritance". He resolutely set out to overcome the "Weismann's barrier", collecting information and carrying out experiments. In particular, he tried to prove the possibility of inheriting immunity. It has long been known that people need vaccinations regardless of whether their parents got them or not but this fact did not stop Steele. He paid attention to another phenomenon concerning immunity: tolerance, or absence of an immune response. It was studied successfully by Peter Medawar and his colleagues in the 1950s. The essence of the experiments conducted in Medawar's laboratory of was as follows: embryos of mice were injected spleen cells from other mice. After the recipient mice reached adulthood, they were grafted a patch of the donor's skin. The graft took, while grafts taken from non-donor mice were rejected. Thus, at the early stages of individual development, when the immune system is yet unformed, tolerance to a particular donor may be induced.

Steele decided to prove that this kind of tolerance was inherited. He conducted a series of experiments using Medawar's methods and reported that almost a half of the offspring in the first generation of experimental mice showed an increased tolerance to the cells to which the parents had been tolerant. In other words, the acquired character, tolerance, was inherited. Medawar and other researchers attempted to reproduce Steele's experiments but failed. A discussion ensued, which strongly resembled the arguments about Kammerer's activity. (In fact, Steele sometimes mentioned Kammerer in the same vein as Lysenko supporters did.)

The consequences of this story were not tragic but nevertheless unpleasant. Steele had great difficulties in finding a position. Even an influential philosopher **Karl Raimund Popper** (1902–1994) could not help, though he approved of Steele's assault of the Weismann's barrier and actively participated in the heated discussions. Steele, in turn, tried to implement Popper's argumentation in scientific research claiming that since his own hypothesis had never been discredited, it had, according to Popper, a right to exist. Apparently, other researchers found this argumentation unconvincing (Steele et al. 1999; Aronova 2003). It turns out that missing an important point at the stage of stating the problem may lead to a long-term deadlock. An attempt to combine Lamarck's philosophy with immunology failed. This is partly due to the fact that neither immunology nor genetics existed in Lamarck's times. A careful analysis of the old theory was in order, and Steel and the others did not manage that. The slogan *What if Lamarck was right?* was patently insufficient.

"Lamarckian heredity" was mentioned in some other relatively modern studies. So, Eva Jablonka and Marion Lamb suggested that epigenetic changes of chromatin could be inherited and influence evolution (Jablonka and Lamb, 1989, 1995). Other such mentions were usually made in studies of unicellular organisms where "adaptive mutations" and other instances of "inheritance of acquired characteristics" were reported. It was found, in particular, that beneficial mutations emerged in strains of yeast and bacteria more frequentely than could be expected in case of random variability. These data were actively discussed in the 1990s (Cairns and Foster 1991; Landmann 1991; Hall 1992; Lenski and Mittler 1993; Shapiro 1995 etc.). One of the main conclusions was as follows: "Acquired characteristics of the extranucleic, epinucleic, or nucleic types may become heritable in single-cell systems and some of these heritable changes may be adaptive. Apparently, however, only characters of the nucleic type can pass the germline of multicellular organisms." (Landmann 1991, p. 17). In a way, these studies closed the question about the inheritance of acquired characteristics. At the same time, they were initially far from "Lamarckian discussions" and never attempted to solve the problem about the direct action of environment upon evolution raised by the authors of the past. Other studies are more relevant to this problem, namely, those in which the results of the environmental impact upon variation were observed irrespectively of the mechanisms of heredity, especially, studies on differences between similar organisms grown under different conditions.

Investigations by an American zoologist **Francis Bertody Sumner** (1874–1945) are especially interesting in this respect. He reared mice under cold and warm conditions for a long time. Mice reared in the cold became larger than their parents and had many other tiny differences from them (other than those in mice bred under warm conditions). Mice that grew in the cold differed from their parents but did so statistically, that is, demonstrated a broad range of variation, with some kinds being similar to the parents (Sumner 1915). Both Lamarckians and their opponents considered these data as though the offspring inherited the acquired characters but could acquire new ones under a new influence. However, in relation to the changed parents the offspring did experience a new influence: "not-cold" conditions.

Neither Lamarckians nor their opponents could interpret these data convincingly, and both warring factions used them to support their views (Sakharov 1952; Provine 1979). Sumner did not pledge himself to any of the parties but patiently continued research on other objects such as deer mice (*Peromyscus*) and fish. The new results were even more difficult to interpret. By the time they were obtained, Lamarckism lost popularity and Sumner's results were used to support modern Darwinism though it is unlikely that the author would have agreed with this. Sumner proved the hereditary nature of the manifestations of geographic variability in deer mice populations but judged, quite correctly, that this finding did not explain how they originated. Moreover, in the process of laboratory breeding, different races of deer mice changed in an abstruse manner. Contrary to expectations, there were no signs of different races converging under the influence of identical conditions. Long breeding resulted in an increased diversity, the emergence of "mutants" and other "anomalies", some of which were not inherited at all. Laboratory deer mice differed from the wild ones in size and body proportions. This may be due to the fact that a much greater proportion of mice survived in the laboratory than in the natural environment and rare kinds of variation occurred much more often under the laboratory conditions. However, Sumner considered these kinds as anomalies due to breeding conditions and tried to achieve the stable reproduction of perfectly normal *Peromyscus* (Sumner 1926, 1932; Provine 1979).

Sumner's observations on deer mice in nature also yielded conflicting results in respect of Lamarckism and Darwinism. A number of facts testified in favour of the hypothesis about a direct effect of humidity on colour. Mice living under dry conditions were lighter than mice living in wet climate, that is, Gloger's rule was observed. Deer mice are nocturnal, so coloration is unlikely to have an adaptive significance. Sumner noted that the rule also worked in some other animals that do not need cryptic coloration such as skunks and bats. In addition, Sumner found that a colony of dark coloured *Peromyscus* inhabited an almost isolated reef with white sand and apparently had little opportunity to mix with the general population. This fact indicated the absence of strong competition on the basis of colour adaptation among these animals. Opposite cases were also discovered. Dark-coloured *Peromyscus crinitus* occurred on a dark lava ground under dry conditions. Conversely, the most light-coloured *Peromyscus polionotus* occurred on a sandy beach in Florida under wet conditions, its colouration fully coinciding with the colour of sand. These data appeared to testify in favour of natural selection and adaptation. Finally, some data disagreed with both Lamarckian and Darwinian explanations. Dark-coloured *Peromyscus crinitus* occurred in large numbers not only on dark lava ground but also in surrounding sandy deserts, and vice versa, many light-coloured mice typical of deserts were found on a dark ground. Sumner made spectacular photos illustrating a complete mismatch of an animal's colouration and its environment (Sumner and Swarth 1924).

Generally, the "forgotten" works of Sumner raise more questions than they give answers. His experiments (as well as many others) prove only that organisms change in the course of generations in any case and that they always react to changes in the environment. Leaving aside misunderstandings concerning Lamarck and Lamarckism, the very phrase "inheritance of the acquired characteristics" is incorrect. It implies that one generation acquires characters and the next one inherits or does not inherit them but does not acquire new ones. To remove this absurdity, one should discuss not the "inheritance of acquired characteristics" but the direct impact of environmental change on organisms and its role in evolution. No one doubts that it can be significant. For example, some crustaceans may change beyond recognition as a result of changes in salinity. Their modifications were described as different species or even genera (Schmankewitsch 1877). This phenomenon can be considered as an evolutionary event even if it is not directly associated with a change in genes. Let us imagine that the crustaceans changed once under the influence of changed water salinity and then the salinity did not change in the opposite direction for millions of years. During this time the genes of crustaceans would change anyway if only because genes cannot reproduce their exact copies for an indefinitely long time. If millions of years later the salinity would revert to the original state, it is unlikely that the organisms would. So, spontaneous evolution related with the direct impact of the environment would take place over long periods of time. This is what Lamarck claimed.

Let us consider another example agreeing with such a Lamarckian description of the evolutionary process. Marine invertebrates basically have two patterns of reproduction: by planktotrophic larvae or by lecithotrophic larvae. In the first case, an animal produces eggs poor in yolk, which grow into larvae which provide for themselves until when they can transform into an adult organism. In the second case, an animal produces eggs rich in yolk and the larvae can start metamorphosis very soon, without additional feeding. Both options are expedient as both have advantages and disadvantages. Lecithotrophic larvae have an easier life but planktotrophic larvae have better chances of dispersal and finding a new suitable habitat. The survival rate of lecithotrophic larvae may be higher but the parents of planktotrophic larvae compensate for this by producing more numerous offspring.

Different degrees of lecithotrophy are known. Lecithotrophic larvae may have a reduced or a completely absent digestive system. The transition from one type of larvae to another may be associated with an intensity of feeding by parental organisms. If they are well nourished, they are in a better position to provide their offspring with food, i.e. to generate lecithotrophic larvae. On the contrary, if the parents eat poorly, they can let their offspring fend for themselves. The differences in this respect can be manifested as a kind of variation (for example, sea urchins and some polychaetes form different types of larvae under different conditions). However, a similar phenomenon can be seen in the evolution of large groups over millions of years. Cheilostome bryozoans experienced several bursts of diversity in the course of their evolution, and one of them was associated with the emergence of lecithotrophy. This event coincided with a change of conditions in the ocean, when much more food became available. An abundance of food persisted for millions of years, and bryozoans have continued to evolve as animals with several types of embryos and larvae (Ostrovsky 2013). We may conclude from this that cheilostome bryozoans evolved under the direct influence of the environment, and this conclusion would be neither mystical nor unscientific. This means that pronouncements about the direct impact of the environment upon evolution, which can be found in evolutionary treatises of the past, are not always anachronistic.

Thus, accusations of Lamarckism piled on the supporters of orthogenesis are hardly deserved. The problem of inheritance of acquired characteristics can easily be dismissed from the scientific agenda. Speculations on this topic are more similar to ancient myths than to Lamarck. As we know, when Phaeton failed to control the sun chariot, the Sun came perilously close to the Earth. The Sahara desert was formed, and the people of Ethiopia became black. Afterwards the order of the Sun movements was restored, but Ethiopians remained black. In other words, one generation had acquired a new feature (black skin) and the subsequent generations inherited it, being unable to acquire a new character under the impact of normal temperature.

References

Aronova EA (2003) Lamarckism and immunology in the second half of the 20th century: an analysis of scientific discussions (Lamarkizm i immunologiya vo vtoroy polovine XX v.: analiz nauchnykh diskussiy). Cand Sci (Biol) Dissertation. IIET RAN, Moscow

Bonnier G (1907) The world of plants (Le monde végétal). Ernest Flammarion, Paris

Cairns J, Foster PL (1991) Adaptive reversion of a frameshift mutation in *Escherichia coli*. Genetics 128:695–701

Gaysinovich AE (1988) Origin and development of genetics (Zarozhdenie i razvitie genetiki). Nauka, Moscow

Gliboff S (2006) The case of Paul Kammerer: evolution and experimentation in early 20th century. J Hist Biol 39(3):525–563

Hall BG (1992) Selection-induced mutations occur in yeast. Proc Natl Acad Sci U S A 89:4300–4303

Houssay F (1907) Experimental variations; studies on six generations of carnivorous chicken (Variations expérimentales; études sur six générations de poules carnivores). Arch zool expér et gén Sér 4:137–332

Jablonka E, Lamb MJ (1989) The inheritance of acquired epigenetic variations. J Theor Biol 139:69–83

Jablonka E, Lamb MJ (1995) Epigenetic inheritance and evolution: the lamarckian dimension. Oxford University Press, Oxford

Lamarck JB (1809) cited in Lamarck (1914) English translation by Hugh Samuel Roger Elliott

Landmann OE (1991) The inheritance of acquired characteristics. Ann rev genet 25:1–20

Lenski RE, Mittler JE (1993) The directed mutation controversy and neo-Darwinism. Science 259(5092):188–194

Olsson L (2003) Resistance to Darwinism in Sweden: idealistic morphology, Lamarckism, the idea of constancy of species and antiselectionism (Soprotivlenie darvinizmu v Shvetsii: idealisticheskaya morfologiya, lamarkizm, ideya postoyanstva vidov i antiselektsionizm). In: in the shadow of Darwinism. Alternative theories of evolution in the 20th century (V teni darvinizma. Al'ternativnye teorii evolyutsii v XX veke). Yasnyy den', St Petersburg, pp 82–94

Ostrovsky AN (2013) Evolution of the sexual reproduction in marine invertebrates: example of gymnolaemate Bryozoa. Springer Verlag

Provine WB (1979) Francis B Sumner and the evolutionary synthesis. In: Studies in the history of biology, vol 3. John Hopkins University press, Baltimore/London, pp 211–249

Sakharov PP (1952) Inheritance of acquired properties (Nasledovanie priobretaemykh svoystv). Sovremennaya nauka, Moscow

Schmankewitsch W (1877) To the knowledge of the influences of life conditions on the organisation of animals (Zur Kenntnis der Einflusses des Lebensbedingungen auf die Organization der Tiere). Z Wiss Zool 29:429–494

Shapiro J (1995) Adaptive mutation: who is really in the garden? Science 268:373–374

Steele EJ, Lindley RA, Blanden RV (1999) Lamarck's signature. How retrogenes are changing Darwin's natural selection paradigm. Perseus books, Reading, Massachusetts

Sumner FB (1915) Some studies of environmental influence, heredity, correlation and growth in the white mouse. J Exp Zool 18:325–432

Sumner FB (1926) An analysis of geographic variation of the Peromyscus group from Florida and Alabama. J Mammology 7(3):149–184

Sumner FB (1932) Genetic, distributional and evolutionary studies of the subspecies of deer-mice (Peromyscus). Bibliographia genetica 5(9):1–104

Sumner FB, Swarth HS (1924) The supposed effects of the color tone of the background upon the coat color of mammals. J Mammology 5(2):81–113

Weismann A (1892) Essays on inheritance and related biological questions (Aufsätze über Vererbung und Verwandte biologischen Fragen). Fischer, Jena

Chapter 14
Mystics or Unrealised Synthesis?

Abstract Orthogenesis has often been accused of "mystics", "vitalism", "teleology" etc. but these accusations were little more than insults used in polemics on evolutionary issues. The main reason why orthogenesis is inconvenient and unpopular is that it places unsolved issues at the centre of the evolutionary concept. The complexity of biological phenomena, which cannot be explained in their entirety by the modern synthesis, keeps attracting new researchers.

Keywords Vitalism · Teleology · Synthesis · Mystics · Unsolved problems of evolutionary biology

Juan Antonio Llorente (1756–1823), an inquisitor and a historian, remarked in his *Critical history of the Spanish Inquisition* that in the thirteenth century "the taste for allegorical interpretation of the Scripture prevailed with time so that its literal sense was as good as ignored." (Llorente 1870, p. 34). The history of Inquisition comes to mind because the arguments of evolutionists are sometimes similar to those used by religious leaders. Just as interpreters of the Scripture denounced new ideas, the participants of debates on directed evolution often engaged in philosophical interpretation of terms and concepts, denouncing each other's fallacies—in fact, accusing each other of a lack of intelligence.

A special analysis was necessary in order to identify the major scientific arguments in the controversy between Darwinism and orthogenesis because they were often hidden in denunciations and verbal tricks. Many critics of orthogenesis tried to show that its supporters were not honest researchers but evildoers aiming to destroy Darwinism with the help of a new device. This approach was especially prominent in the Soviet Russia. The communist ideology gave ample opportunities to denounce anything even remotely suggestive of a deviation from materialism. It was to deflect such criticism that articles about orthogenesis (or nomogenesis) contain numerous references to dialectical materialism. Strange as it may seem, criticism of directed evolution in the West had much in common with what was presented under the banner of Marxism. The opponents of Darwinism were exposed as people who do not understand the obvious. Exaggeration, distortion, ridicule, reduction to

I. Popov, *Orthogenesis versus Darwinism*,
https://doi.org/10.1007/978-3-319-95144-7_14

absurdity—all these tricks were commonly used in argumentation against directed evolution.

Alexander V. Nikolsky (1958–1942), a critics of nomogenesis, discussing the phenomenon of phylogenetic acceleration mentioned by Berg noted that a human embryo is covered with hair. *What would the future man be similar to? An ape or a devil?* he asked (Nikolsky 1928, p. 57). Nikolsky compared nomogenesis to a strange, capricious, treacherous and cunning creature and remarked that Sobolev and Berg lacked intellect. The architects of the evolutionary synthesis, Mayr and Simpson, made statements in a similar style. Simpson claimed that much of the directionality of evolution resulted from the directionality of minds of scientists rather than the presence of such a trend in nature (Simpson 1944). Mayr accused the opponents of Darwinism of an astonishing ignorance of the basics of genetics and the modern literature and remarked that to refute them would be a waste of time (Mayr 1963).

Some orthogeneticists gave tit for tat. Lyubishchev responded to Mayr's statement by saying that "a conscious ignorance of unwanted facts shown by E. Mayr reveals him as a dogmatist blindly believing in his postulates rather than a sound scientist" (Lyubishchev 1973, p. 49). In general, however, since the supporters of orthogenesis were less numerous than their opponents, in terms of quantity fewer of them used derogatory arguments.

Unfortunately, such "argumentation" has recently appeared even in reputable journals. In a review of the book by Otto H. Schindewolf, S. Conway Morris piled rude remarks on his work as well as his person ("a deeply muddled book", "half-baked ideas", a "travesty of what is observed", "this book should be jettisoned into the dustbin of history") (Conway Morris 1994). The terms "decadence", "abnormality" and "degeneration", which Schindewolf used to describe of one of his concept, were considered by S. Conway Morris as a manifestation of Nazism ("just the pejoratives that the Nazis employed in their monstrous programme"—Conway Morris 1994). This kind of criticism is evidently unscientific, and would not have deserved mention, had it not appeared an academic milieu.

Some other non-biological arguments, those related to philosophy and history, were more apposite. Critics often discerned in orthogenesis erroneous philosophical prerequisites or some harmful influence. **Boris M. Kozo-Polyansky** (1890–1957) suggested that nomogenesis resulted from Berg's mysticism, which, in turn, was caused by the hardships of the war and the revolution (Kozo-Polyansky 1923). Nomogenesis and some other concepts of directed evolution indeed sometimes contained philosophical and historical excursions. The cause and effect relationship is not quite clear: either philosophy and history of science had an impact the theory of evolution or the concept was first formed on the basis of biological research and later illustrated by philosophical and historical examples. Whatever the case, philosophical digressions did not clarify the discussions. Darwinists accused orthogenesis of teleology, idealism, "a rotten idea or purposefulness", metaphysics etc. Their opponents, who could philosophise equally well, accused Darwinists of the same sins. The words "vitalism" and "teleology" were mentioned especially often in this context. As is always the case with philosophical terms, their clear definitions were difficult to find, and so the disputants could easily treat these terms as it suited them. The concepts of both vitalism and teleology are so vague that they can be found in both orthogenesis and Darwinism.

The notion of vitalism is usually mentioned in the literature of evolutionary biology as a negative characteristics and, moreover, a self-explanatory one. There are accusations of vitalism but vitalism itself seems not to be there at all. It is difficult to find a scientist who openly proclaimed himself a vitalist. One of the few was **Hans Adolf Eduard Driesch** (1867–1941), the author of a "system of vitalism". His book (Driesch 1905) is best suited for the role of the primary source of the concept of vitalism in the context of the controversy between orthogenesis and Darwinism because it was published at the peak of popularity of ideas of directed evolution. Nevertheless Driesch was rarely cited by orthogeneticists. Out of all the authors mentioned above, only Karl Beurlen did that. The publication of the Russian translation of Driesch's book provoked some interesting comments elucidating the nature of the myth of vitalism. **Nikolay K. Koltsov** (1872–1940), one of the founders of genetics in Russia, wrote: "Russian biologists should rejoice in the publication of the Russian translation of the book by Driesch, and the opponents of vitalism should perhaps rejoice even more than the few supporters of this direction. Until now, it was impossible for Russian readers to become more or less fully acquainted with the views of vitalists." (Koltsov 1914, p. 1527). This shows that until 1915 vitalism was poorly known in Russia. After that, Russian biologists could become familiar with it owing to the Driesch's book yet none of the papers on nomogenesis mentions vitalism. It would seem that in such a case there could be neither opponents nor proponents of vitalism. Nevertheless, they existed. In the West, it seems, supporters and opponents of vitalism appeared in the same mystical manner. This term was often used but none of the papers cited in the preceding chapters contains a reference to its primary source.

Let us try to find out what "vitalism" means. It is difficult to establish who first pronounced the mysterious word. A term closest to vitalism seems to have been used in the early 19th century by **Gottfried Reinhold Treviranus** (1776–1837), who mentioned *vis vitalis* in his six-volume treatise *Biology and Philosophy of Nature* (1802–1822). In the spirit of his time, Treviranus postulated the existence of a "vital force" present in the animate world and absent in the inanimate one. At that time, few seemed to doubt its existence. Treviranus discussed various aspects of its relationship with "matter" and other fundamental principles (Driesch 1905). The idea of the vital force can be found in the works of most thinkers and researchers dealing with biology before the nineteenth century. Some of them, such as Jan Baptist van Helmont (1580–1644), referred to the biblical myths but others such as Caspar Friedrich Wolff (1733–1794) tried to investigate the vital force by analogy with electrical, chemical and other forces. These "vitalists" were engaged in polemics with mechanists, who tried to study living organisms by analogy with machines. Vitalists were apparently more successful because information about the complexity of biological processes was gradually emerging. Fertilisation, ontogeny, regeneration etc. could hardly be explained in mechanical terms. The mechanical approach lost popularity, and it became meaningless to actively develop an opposite viewpoint. According to Driesch, in the 19th century the old vitalism no longer had enemies, and expired peacefully.

New enemies of vitalism appeared with the birth of Darwin's theory. According to vitalists, Darwinism ignores the complexity of the phenomena involved in the transformation of the living world. It is similar to the statement that a house can build itself of stones rolling down the mountain slope (Driesch 1905). This is neither an exaggeration nor a distortion because Darwin did use a similar analogy comparing natural selection with an architect who has to build a majestic building with crude stones and yet fulfils this task successfully (Darwin 1859). Obviously, such an idea is directly opposed to vitalism.

A new enemy has brought to life a new vitalism. Emerging at the time when orthogenesis was being formed, it was different from the old one. Hans Driesch defined its essence as follows: "The main question of Vitalism is not whether the processes of life can properly be called purposive: it is rather the question if the purposiveness in those processes is the result of a special constellation of factors known already to the sciences of the inorganic, or if it is the result of an autonomy peculiar to the processes themselves." (Driesch 1914, p. 1). At the same time, the concept of purposiveness was used descriptively to denote that there was some meaning and order in biological processes. Approximately at the same time, in 1907, Henri Bergson observed that vitalism was very flexible and could not be denied completely. According to him, the aim of researchers of the biological fundamentals was to describe the scope of the specificity of the living world as compared to the non-living ones (Bergson 1998).

Vitalists of the late 19th—early 20th century did not consider orthogeneticists as brothers-in-arms though believed they were better than Darwinists. From the viewpoint of vitalism, the works of orthogeneticists were valuable inasmuch as they contributed to the criticism of the natural selection theory. Contradictions between orthogenesis and vitalism are evident: while vitalists stressed the uniqueness of life compared to inorganic world, orthogeneticists were more inclined to an opposite extreme, analysing biological material by analogy with chemical and physical processes. From the vitalistic point of view, this was reductionism, mechanism, materialism and so on. They believed that not only orthogenesis but also the theory of evolution in general was a not-too-valuable speculation. Therefore, vitalism cannot be identified with orthogenesis. Nikolay N. Gaydukov (1925) noticed this, pointing that accusations of vitalism directed at orthogeneticists were meaningless. On the contrary, *vitalists* used the works of Nägeli (and others) but Nägeli and his followers were not vitalists. Gaydukov compared polemicists arguing about evolution with lawyers who, defending criminals, juggle with facts to achieve their purpose.

Later the understanding of vitalism in biology underwent an amazing metamorphosis. The researchers trying to bring biology closer to exact sciences acquired a reputation for vitalism, while the "official" ideology, which was supposed to be strictly materialistic, proclaimed the uniqueness of biological processes. The idea about a special place of biology in the rank of sciences is, perhaps understandably, quite popular among biologists. It is thought to be so unique that the mere idea of any biological laws and formulas is laughable. Until very recently, some biological theorists claimed emotionally that living organisms could not be measured by the standards of physics and chemistry, that they were too complex, that biology was a kind of art etc. In particular, such ideas were promoted in one of the last books by Ernst Mayr (1997). He argued that biology knew no laws and was essentially different from sciences dealing with inanimate nature. If anything, it was similar to history. Almost the same ideas were voiced in other countries independently of Mayr (e.g. Krasylov 1984).

Modern theorists are even more resolute in such comments than vitalists. Their statements fit the outline of vitalism made by Driesch (1914). Driesch studied the history of vitalism starting from Aristotle. He rejected the use of biblical myths in this context and wrote condescendingly about the researchers of the 16th–18th centuries, who did this. He also criticised the vitalism of Treviranus and his followers, that "really dogmatic Vitalism, i.e. Vitalism whose proof is no longer thought necessary, the question being rather how it is to be presented." (Driesch 1914, p. 100). Dealing with the contemporary stage of vitalism, Driesch presented its "system". He wrote about the specificity of biological processes with some caution, aiming to explore if there were any processes of life that could not be reduced to purely physical or chemical ones and therefore had to be regarded as "autonomous", that is, subordinate to their own laws (Driesch 1914). Modern theorists consider this century-old question as solved, and the answer is that the life processes are autonomous. At the same time, as we have seen, some modern advocates of orthogenesis support mechanicism even more strongly than their predecessors. Frankfurt morphologists, who see the sources of their "school" in the writings of Descartes (Levit et al. 2002), actively develop the idea of similarity between machines and organisms, which is a direct opposite of vitalism.

This means that mechanism and vitalism have long swapped places in biology. The discussion of these philosophical concepts in the context of evolutionary theory is now counterproductive and not unlike sailing between Scylla and Charybdis (Berdyshev and Siplivinsky 1961; Gutmann and Weingarten 1992; Havel 1997).

The authors of concepts of directed evolution did sometimes mention teleology, mysticism, idealism etc. but not as something they wanted to justify. They found these sins in the works of others, including each other. For example, Kölliker accused the theory of natural selection of teleology because it proceeded from the assumption that all details of the organism's structure had been created for its benefit. Kölliker, similarly to other orthogeneticists, tried to prove the opposite. Nägeli argued about this both with Darwin and Kölliker in 1860. According to him, at least some features of organisms were indeed adaptive. Later, however, he paid more attention to the criticism of the adaptation paradigm. Nägeli himself was also accused of mysticism. Strangely, one of the accusers was Theodor Eimer, whose ideas were close to Nägeli's. Eimer called Nägeli's doctrine mystical. According to him, a direct action of environment was a material factor, while spontaneous transformation was mysticism. Eimer, in turn, was also accused of mysticism. Charles Whitman, for instance, wrote that Eimer's idea were a new teleology (Whitman 1919). At the same time, after his long-term studies Whitman came to the conclusion that orthogenesis really existed, though he considered himself neither a follower of Eimer nor a teleologist.

The inconsistency of such comments betrays them as emotional assessments, which do not belong to the sphere of scientific debates. Many orthogeneticists indeed dabbled with non-materialistic philosophy and/or religion but their writings on evolution were usually presented in a proper scientific style. The most notable exception is the work of Teilhard de Chardin and his followers. Works by some German authors such as Dacqué and Beurlen can also be considered as such exceptions. The work of Karl E. von Baer, who wrote about the purposefulness of development, might seem yet another exception but even he tried to distance himself from non-materialistic philosophy, mysticism or teleology. Discussing development, Baer carefully analysed intricate semantic differences between the German words *Zweck* and *Ziel*, which in ordinary use are quite synonymous and can be translated into English as “goal” or “aim”. According to Baer, *Zweck* was a goal put consciously and freely, while *Ziel* was the end of the action in a certain direction or a predetermined desired result of the process. Baer described natural processes using the word *Ziel*, emphasising its descriptive sense and opposing vitalistic speculations (Blyakher 1955). This means that even his work cannot be unambiguously classified as unscientific teleology.

It is hardly productive to instigate a search for vitalism or teleology in evolutionary biology. The denial of these ideas leads to mechanicism, whose reputation in biology is also poor. At the same time, mechanicism, similarly to vitalism, has adapted to new conditions with the progress of evolutionary biology. The famous work *L’évolution créatrice (Creative Evolution)* by Henri Bergson is a demonstrative example. Bergson may be said to have been cautiously sailing between Scylla and Charybdis of mechanicism and vitalism, indeed, between all the scyllae and charybdi waiting for an evolutionist. Sometimes he followed Descartes: “That life is a kind of mechanism I cordially agree.” (Bergson 1998, p. 32, translated into English by Arthur Mitchell). At the same time, he wrote warmly about vitalism. The following is especially noteworthy: “...the ‘vital principle’ may indeed not explain much, but it is at least a sort of label affixed to our ignorance, so as to remind us of this occasionally, while mechanism invites us to ignore that ignorance.” (Ibid., p. 44–45). Bergson drew attention to the fact that our ignorance is, in a certain sense, only progressing. It would seem that the advances in the study of tissues and cells would “decompose” life into inorganic components but, quite on the contrary, the gap between the animate and the non-animate world only deepens. It is hardly clearer now than before how living beings could arise from a non-living matter. Any ambiguity in biological phenomena is suggestive of vitalism.

Therefore, traces of vitalism in orthogenesis are not an evidence of mysticism or an unscientific approach and not a reason to exclude its supporters from the academic community. Moreover, almost all the authors of the concepts of directed evolution discussed in this book were very productive researchers. Some of them such as Berg, Vandel, Nägeli and Kölliker were classics, founders of new research trends in biology. Mystics are not like that. Pseudoscientists and undisguised mystics are usually content with superficial knowledge.

Thus, there are no good reasons to accuse a scientist of mysticism barely because he criticises the theory of natural selection and approves of the idea of directed evolution. The ideological, philosophical and other non-scientific arguments obscure the valuable aspects of orthogenesis. The modern synthesis, that is, the incorporation of the greatest possible amount of materials and trends into a theory, is considered as the greatest achievement of evolutionary biology and the main merit of Darwinism. Yet orthogenesis also offers possibilities for a synthesis.

We have seen that orthogenesis was born from a bizarre union of different, often opposite trends in evolutionary biology. The concepts of directed evolution were developed by followers of Lamarckism, saltationism and Neo-Darwinism. These trends have often opposed each other but their contradictions seem insignificant in the light of orthogenesis, which focuses on the regularities of variability, that is, a predisposition to vary in certain directions or, in other words, restrictions on possible changes. If viewed in this light, the rates of evolution or the details of variability mechanisms seem secondary issues, not crucially important for the fundamentals of the evolutionary theory.

Moreover, some input into orthogenesis was made by theoreticians who were not evolutionists. It was because this concept made it possible to overcome the conflict of the typological and the evolutionary approach in the analysis of the diversity of living organisms. This issue was important at the time when orthogenesis was formed. Old morphology implied the study of types and their relationships. The major theoreticians of pre-Darwinian biology were more concerned with similarities and differences between types rather than with the vicissitudes of their evolution. In post-Darwinian times, increasingly more attention has been paid to generalizations based on the evolutionary idea. The diversity of organisms was considered as a result of the explosive growth of phylogenetic branches in all directions without restrictions, and the analysis of the organisms' features focussed on their adaptive significance. Idealistic morphology, that is, the search for typological affinities or similarities irrespectively of evolutionary ways and factors, found itself at the periphery of biology. Orthogenesis could combine both approaches. On the one hand, it implied careful studies of phylogenetic trends but, on the other hand, necessitated typological studies since the trends were seen as leading to something strictly ordered.

It is interesting that the search for the periodic law in chemistry was based on similar considerations. Dmitry Mendeleev looked for "natural groups" of elements and their compounds. In the language of biology, he looked for a typological affinity or similarity as idealistic morphologists did. At the same time, Mendeleev analysed the evolution of elements, that is, described various series of smooth changes in the properties of elements and their compounds (Kedrov 1970; Dmitriev 2001).

Finally, orthogenesis pointed a way towards the connection between biology and other sciences. Almost all authors of the concepts of directed evolution postulated the existence of biological laws similar to physical and chemical one. Orthogenesis offers a tempting possibility of describing evolution by analogy with the processes of movement studied by physics and of composing systems of organisms by analogy with the periodic system of elements.

Yet in spite of all these promising vistas, orthogenesis remains in the shadow of the modern synthesis. It has made little progress, and no evolutionary synthesis has been achieved on its basis. Perhaps this is partly because the new always makes its way slowly. Humans could not find a scientific explanation of natural phenomena for thousands of years, and absurd superstitions persist even in a civilised society. An explanation of this was given in *The Golden Bough: A Study in Magic and Religion* by James George Frazer (1854–1941), and it seems appropriate to cite a passage from it.

"The reader may well be tempted to ask, how was it that intelligent men did not sooner detect the fallacy of magic? How could they continue to cherish expectations that were invariably doomed to disappointment? With what heart persist in playing venerable antics that led to nothing, and mumbling solemn balderdash that remained without effect? Why cling to beliefs which were so flatly contradicted by experience? How dare to repeat experiments that had failed so often? The answer seems to be that the fallacy was far from easy to detect, the failure by no means obvious, since in many, perhaps in most cases, the desired event did actually follow, at a longer or shorter interval, the performance of the rite which was designed to bring it about; and a mind of more than common acuteness was needed to perceive that, even in these cases, the rite was not necessarily the cause of the event. A ceremony intended to make the wind blow or the rain fall, or to work the death of an enemy, will always be followed, sooner or later, by the occurrence it is meant to bring to pass; and primitive man may be excused for regarding the occurrence as a direct result of the ceremony, and the best possible proof of its efficacy. Similarly, rites observed in the morning to help the sun to rise, and in spring to wake the dreaming earth from her winter sleep, will invariably appear to be crowned with success, at least within the temperate zones; for in these regions the sun lights his golden lamp in the east every morning, and year by year the vernal earth decks herself afresh with a rich mantle of green. Hence the practical savage, with his conservative instincts, might well turn a deaf ear to the subtleties of the theoretical doubter, the philosophic radical, who presumed to hint that sunrise and spring might not, after all, be direct consequences of the punctual performance of certain daily or yearly ceremonies, and that the sun might perhaps continue to rise and trees to blossom though the ceremonies were occasionally intermitted, or even discontinued altogether. These sceptical doubts would naturally be repelled by the other with scorn and indignation as airy reveries subversive of the faith and manifestly contradicted by experience. "Can anything be plainer," he might say, "than that I light my twopenny candle on earth and that the sun then kindles his great fire in heaven? I should be glad to know whether, when I have put on my green robe in spring, the trees do not afterwards do the same? These are facts patent to everybody, and on them I take my stand. I am a plain practical man, not one of your theorists and splitters of hairs and choppers of logic. Theories and speculation and all that may be very well in their way, and I have not the least objection to your indulging in them, provided, of course, you do not put them in practice. But give me leave to stick to facts; then I know where I am." The fallacy of this reasoning is obvious to us, because it happens to deal with facts about which we have long made up our minds. But let an argument of precisely the same calibre be applied to matters which are still under debate, and it may be questioned whether a British audience would not applaud it as sound, and esteem the speaker who used it a safe man—not brilliant or showy, perhaps, but thoroughly sensible and hard-headed. If such reasonings could pass muster among ourselves, need we wonder that they long escaped detection by the savage?" (Frazer 1922, Part IV. p. 16).

What if orthogenesis is an alternative to the "magic" of the theory of natural selection? What if organisms can evolve without natural or artificial selection? All too often Darwinists behave like Frazer's "safe man", relying, in their criticism of alternative theories, on "the opinion of the majority" and the "method of exclusion".

Unimpressive progress of orthogenesis is also due to the estrangement of its supporters. Orthogenesis is nearly 150 years old. No less than 20 concepts of directed evolution have been created, often independently of each other, during this period (Table 14.1). Some of them really incorporated material that was new at the time (e.g. biochemical evolution) or presented new ideas. Others, however, merely clad an old concept in a new dress. Researchers often believe that it is better to take a fresh look on a complex phenomenon, without being encumbered by tradition or authority. Yet human thinking is extremely stereotypical, and "new" ideas are likely to follow a trodden path. The result is a waste of effort on repetition instead of creating something new.

Terminology and language of publications on directed evolution are worthy of special attention in this respect. While it is clear that an estrangement may arise from a lack of knowledge of foreign languages, many authors of orthogenetic concepts wrote in a language of their own, which was unknown to the others. It would seem that by the time the term "orthogenesis" was more than a hundred years old, the authors might have written about its aspects, mechanisms, manifestations and so on. However, for some reason they preferred to coin new terminology. As a result, we now have to perform multiple comparisons such as "autogenesis = nomogenesis = autoevolution ≈ typogenesis", "the orthogenesis of Haacke ≠ the orthogenesis of Eimer", "rectigradation ≈ orthogenesis" etc. Only the most important terms related to directed evolution were mentioned in this book. To incorporate all of them, a voluminous glossary would be needed. The compilation of such a glossary might be an aim of a special research but does not seem to be all that necessary: many of the terms have never been used outside the work where they were introduced.

Some authors dealing with the history and theory of evolutionary biology also contributed to the terminological confusion, suggesting new terms related to directed evolution or new wordings of the existing ones. So, Julian Huxley divided orthogenesis into dominant orthogenesis and subsidiary or secondary orthogenesis (Huxley 1942), although the difference between them is eluding. Gould divided orthogenesis into a hard and a soft one (Gould 2002), and both were different from the meaning ascribed to this term in the original sources. According to Gould, hard orthogenesis was the idea of predetermined extinction and aging of taxa (in particular, the concept of Hyatt, who was mostly active before the term orthogenesis was introduced) though Eimer and Haacke, the first authors to suggest this term, had

Table 14.1 Milestones of the history of orthogenesis

Author, years of life	Country	Field of research	Evolutionary concept and/or its main terms	Years of major publications
Carl Wilhelm von Nägeli, 1817–1891	Switzerland, Germany	Botany, microanatomy and physiology of plants	Mechanic-physiological theory of evolution, theory of perfection	1856–1884
Rudolf Albert Kölliker, 1817–1905	Switzerland	Histology, zoology	Heterogenesis, perfection theory	1864–1872
Edward drinker cope, 1840–1897, and his followers	USA	Palaeontology	Bathmism, Bathmogenesis, Neo-Lamarckism	1868–1897
Karl Ernst von Baer, 1792–1876	Russia	Embryology	Purposeful creation	1876
Wilhelm Haake, 1855–1912	Germany	Zoology	Orthogenesis	1893
Theodor Eimer, 1843–1898	Germany	Zoology	Orthogenesis	1897
Daniel Rosa, 1857–1944	Italy	Zoology	Ologenesis	1898–1931
Henry Fairfield Osborn, 1857–1935	USA	Palaeontology	Aristogenesis	1912–1934
Leo S. Berg, 1876–1950	Russia	Ichthyology	Nomogenesis	1922–1926
Dmitry N. Sobolev, 1872–1949	Russia	Palaeontology	Historical biogenetics	1924
William Dickson Lang, 1878–1966	Great Britain	Palaeontology	Orthogenesis	1910–1920s
Hans Przibram, 1874–1944	Austria	Embryology	Apogenesis	1900–1920s
Alphonse Labbé, 1869-?	France	Zoology	Allelogenesis	1920s
André Lwoff, 1902–1994	France	Physiology	Physiological reduction, orthogenesis	1930–1940s
Pierre Teilhard de Chardin, 1881–1955	France	Palaeontology	Orthogenesis, omega principle	1950
Otto Heinrich Schindewolf, 1896–1971	Germany	Palaeontology	Typostrophism	1930s–1950s
Albert Vandel, 1894–1980	France	Zoology	Organicism	1960s
Leon Croizat, 1894–1982	Italy, USA, Venezuela	Botany, Biogeography	Panbiogeography, Biological synthesis	1960s

(continued)

Table 14.1 (continued)

Author, years of life	Country	Field of research	Evolutionary concept and/or its main terms	Years of major publications
Miguel Crusafont-Pairó	Spain	Palaeontology	Orthogenesis, Orthogenesis (with a capital letter)	1950s–1960s
Wolfgang Gutmann (1935–1997) and his followers	Germany	Morphology	Frankfurt school	1970s–1990s
Antonio Lima de Faria	Sweden	Cytogenetics	Autoevolution, biological periodicity	1980s–1990s
Kunio Kawamura	Japan	Chemistry	Individuality, self-organisation, shutayzey	1990–2003
Eduard B. Akhnazarov (1933–2001)	Russia	Economics	Contours of evolution, impact-reflection	2002

used it in an entirely different sense. The phenomenon that Gould referred to as hard orthogenesis had been previously called by other names (organicism and, partly, ologenesis). Soviet critics of nomogenesis also added some new words such as telegenesis, brockism and neonomogenesis. Sometimes the coinage of new terms stemmed from the ignorance of primary sources but it is also found, quite inexplicably, in studies which contain references to primary sources. Whatever the case, it is inacceptable.

Such an estrangement is a significant setback for the progress. Yet a slow development of orthogenesis seems to be mainly due to another reason, which is associated with the argumentation of its supporters and opponents. In fact, there is often no argumentation at all: the idea of directed evolution is merely proclaimed. It is especially typical of the latest concepts. The prospects of progress are doubtful in such a case. Studies where a "new" idea is merely declared lose much in comparison with the works of Darwin, where each claim, and even the very right of the author to suggest hypotheses, is thoroughly substantiated.

When orthogeneticists present arguments, they usually concern phenomena that are equally difficult to explain from a Darwinian or an alternative viewpoint. The stance on these phenomena can be summed up by a traditional ending of biological papers: "further research is necessary to clarify this problem". It seems that orthogenesis places unsolved issues at the centre of the evolutionary concept. In this respect, it is indeed somewhat similar (though not related) to vitalism (at least, in some interpretations). Just as vitalism reminds us of issues which mechanicism invites us to forget, so orthogenesis highlights questions considered as minor ambiguities by the theory of natural selection. This is the main reason why orthogenesis is inconvenient and, therefore, unpopular.

However, the complexity of biological phenomena, which cannot be explained in their entirety by the modern synthesis, keeps attracting new researchers. Orthogenesis is reborn, phoenix-like, again and again. The tenacity of the concepts of directed evolution is illustrated by the fact that they have been expressed by authors who worked in various fields of biology in different countries and at different times.

References

Berdyshev GD, Siplivinsky VN (1961) The first Siberian professor of botany, Korzhinsky. On occasion of centennary of his birth (Pervyy sibirskiy professor botaniki Korzhinskiy. K 100-letiyu so dnya rozhdeniya). Sibirskoe otd. AN SSSR, Novosibirsk

Bergson H (1998) Creative evolution (trans: Mitchell A). Dover publications Inc, Mineola

Blyakher LYa (1955) History of embryology in Russia. 18th–19th century (Istoriya embriologii v Rossii. XVIII–XIX vv). AN SSSR, Moscow

Darwin Ch (1859 [1987]) The Origin of Species. (Reprint of 1st edition 1859) Suffolk: Penguin books

Conway Morris S (1994) Wonderfully, gloriously wrong. Trees 9(10):407–408

Dmitriev IS (2001) A scientific discovery in statu nascendi: the periodic law of D. I. Mendeleev (Nauchnoe otkrytie in statu nascendi: periodicheskiy zakon D. I. Mendeleeva). Voprosy istorii estestvoznaniya i tekhniki 1:31–82

Driesch H (1905) Vitalism as history and as theory (Der Vitalismus als Geschichte und als Lehre). Verlag von Johann Ambrosius Barth, Leipzig.

Driesch H (1914) The history and theory of vitalism (trans: Ogden CK). Macmillan and Co., Limited, London

Frazer JG (1922) The golden bough: a study in magic and religion. Macmillan, New York

Gaydukov NM (1925) On the phylogenetic system of angiosperms in connection with serum diagnosis, complications and convergencies (O filogeneticheskoy sisteme skrytosemennykh v svyazi s serodiagnostikoy, komplikatsiyami i konvergentsiyami). Zapiski Belorusskogo instituta sel'skogo i lesnogo khozyaystva, 8: 67–81

Gould SJ (2002) The structure of evolutionary theory. Harvard University Press, Cambridge

Gutmann M, Weingarten M (1992) Principles of construction morphology and organismic theory of evolution (Grundlagen von Konstruktionsmorphologie und organismischer Evolutionstheorie). Aufsaetze und Reden senck naturf Ges 38:51–68

Havel V (1997) Address to FORUM 2000 Conference. Prague Castle, 4

Huxley JS (1942) Evolution. The modern Synthesis. Allen & Unwin, London

Kedrov BM (1970) The microanatomy of a great discovery (Mikroanatomiya velikogo otkrytiya). Nauka, Moscow

Koltsov NK (1914) Professor Hans Driesch. Vitalism. Its history and system. Authorised translation of professor A.G. Gurvich with addition to the Russian edition of the author and the translator. Nauka publishing house. Moscow, 1915 (Prof. Gans Drish. Vitalizm. Ego istoriya i sistema. Avtorizirovannyy perevod prof. A. G. Gurvicha s dopolneniyami k russkomu izdaniyu avtora i perevodchika. Knigoizdatel'stvo «Nauka». Moskva, 1915). Priroda 12:1528–1530

Kozo-Polyansky BM (1923) The last word of anti-Darwinism. Presentation and critical analysis of the theory of nomogenesis, a new teaching about evolution of organic world (Poslednee slovo antidarvinizma. Izlozhenie i kriticheskiy razbor teorii nomogeneza, novogo ucheniya ob evolyutsii organicheskogo mira). Burevestnik, Krasnodar

Krasilov VA (1984) The theory of evolution: a need for a new synthesis (Teoriya evolyutsii: neobkhodimost' novogo sinteza). In: Krasilov VA (ed) Evolutionary studies. Marcoevolution (Evolyutsionnye issledovaniya Makroevolyutsiya) DVNTs RAN, Vladivostok, pp 4–12

Levit GS, Gudo M, Krumbein WE (2002) Mechanism in the 21th century (Mechanismus im 21. Jahrhundert). In: Hossfeld U, Th J (eds) Die Entstehung biologischer Disciplinen. II VWB, Berlin, pp 97–124

Llorente JA (1870) A critical history of the Spanish inquisition (Historia crítica de la inquisicion de España). Juan Pons, Barcelona https://archive.org/details/historiacrticad01llorgoog
Lyubishchev AA (1973) On the postulates of modern selectogenesis (O postulatakh sovremennogo selektogeneza). In: Vorontsov NN (ed) Problems of evolution (Problemy evolutsii), vol 3. Nauka, Novosibirsk, pp 31–57
Mayr E (1963) Animal species and evolution. Belknap Press of Harvard University Press, Cambridge (Mass)
Mayr E (1997) This is biology: the science of living world. Belknap Press, Cambridge (Mass)
Nikolsky AM (1928) Nomogenesis (Nomogenez). In: Kozo-Polyansky BM (ed) the theory of nomogenesis. A new phase in the development of Russian anti-Darwinism. A collection of critical papers (Teoriya nomogeneza. Novaya faza v razvitii rossiyskogo antidarvinizma. Sbornik kriticheskikh statey). Izdatel'stvo Gosudarstvennogo timirhyazevskogo nauchno-issledovatel'skogo instituta, Moscow, pp 26–62
Simpson GG (1944) Tempo and mode in evolution. Columbia University Press, New York
Whitman ChO (1919) Orthogenetic evolution in pigeons. In: Riddle O (ed) Posthumous works of Charles Otis Whitman. Carnegie Institute, Washington

Chapter 15
Conclusion

Abstract Concepts of directed evolution have been emerging over the entire period of the existence of evolutionary biology, and keep doing so nowadays. At the same time, the main arguments are being repeated again and again. The following evidence has often been interpreted in favour of directed evolution: constraints on variation, purely age-related changes in the evolution of groups of organisms, preadaptations, parallelisms and convergences, maladaptive characters. The mechanisms of these phenomena are equally difficult to explain regardless of whether Darwinism or its alternatives are chosen as the theoretical framework. The supporters of the idea of directed evolution paid special attention to these phenomena and collected abundant material on them, identifying in this way promising research directions in evolutionary biology. Seen in this regard, orthogenesis is a productive, though unrealised trend rather than a dead end of the evolutionary thought.

Keywords Concepts of directed evolution · Arguments · History

The idea of directed evolution is scattered across numerous studies. The milestones of its history are represented by more than 20 concepts: the theory of perfection by Carl Nägeli (1856, 1865, 1884), heterogenesis by Rudolf Kölliker (1864, 1872), bathmism or Neo-Lamarckism by Edward Cope and his colleagues (Cope 1868, 1896), purposeful creation by Karl Baer (1876), orthogenesis by Wilhelm Haacke (1893), orthogenesis by Theodor Eimer (1897), orthogenesis by William Lang (1916, 1919), ologenesis by Daniele Rosa (1931), aristogenesis by Henry Osborn (1934), nomogenesis by Leo Berg (1926), historical biogenetics by Dmitry Sobolev (1924), typostrophism by Otto Schindewolf (1936, 1950), apogenesis by Hans Przibram (1929), allelogenesis by Alphonse Labbé (1924), physiological degradation by André Lwoff (1943), orthogenesis by Emilio Aguirre (1958), orthogenesis by Pierre Teilhard de Chardin (1950, 1961), organicism by Albert Vandel (1964), biological synthesis by Leon Croizat (1958, 1962), Frankfurt School (Gutmann and Weingarten 1992, 1994; Gutmann 1994), autoevolution and biological periodicity by Antonio Lima de Faria (1988, 1995), the concept of individuality by Kunio Kawamura (2002) and the contours of evolution by Eduard Akhnazarov (2002).

I. Popov, *Orthogenesis versus Darwinism*,
https://doi.org/10.1007/978-3-319-95144-7_15

It is tempting to classify concepts of directed evolution in the same way as biologists classify living organisms. Several considerations preclude this. Firstly, one cannot forget, even for a moment, that each object of such a classification would be a scientist or a scientific group backed by a huge amount of material and ideas. A scientist, unlike a taxonomic object, has a right to classify everything including himself. There is also a discouraging historic precedent: Linnaeus, not content with his success in the classification of organisms, composed a system of contemporary botanists but this attempt was derided and parodied to such an extent that he even thought of giving up his studies.

Secondly, it is hardly possible to compose a system of evolutionary concepts that would be useful in historical and theoretical biological studies. Certain rules and traditions have been established in biological taxonomy (and yet, as any biologist knows, there is not a single group which does not require a complete revision). There are no rules and traditions in the "taxonomy of concepts", and any criterion for classifying versions of orthogenesis would be extremely controversial. Therefore, summing up the history of the idea of directed evolution, I will just recall its most remarkable aspects.

1. It is impossible to name the first or even the main author of the concept of directed evolution. In the second half of the 19th century it was presented more or less independently by at least six scientists (Kölliker, Nägeli, Cope, Hyatt, Haacke, Eimer). They all claimed that organisms spontaneously transformed in certain directions in the process of evolution and denied the evolutionary significance of natural selection. Their other views on evolution were quite different. So, there is no primary wording of the concept of directed evolution. We can only infer its essence from the context of the works of several authors.
2. A clear definition of the term "orthogenesis", which is most often used to denote the idea of directed evolution, is also lacking. In its primary wording, it concerned the ability of organisms to vary only in certain directions. However, almost immediately this term was applied to characterise various things such as differences between species, phylogenetic trends and certain mechanisms of evolutionary process. Strictly speaking, it was used incorrectly immediately after it had been suggested because it was intended for the analysis of extremely complex problems. Therefore, it is only with certain reservations that the term "orthogenesis" can be used to denote all concepts of directed evolution.
3. The most noticeable change in the history of the idea of directed evolution concerns the problem of the direct action of the environment on evolution and the evolutionary role of habits and exercises. These ideas are generally considered as Lamarckism. Early theories of the 19th century were usually to some extent associated with them. Later, in the early 20th century, the supporters of orthogenesis either denied the direct influence of the environment on evolution and the effect of exercise or did not emphasise these factors. Finally (after the 1930s), these Lamarckian aspects were abandoned altogether.
4. The demarcation line between orthogenesis and Darwinism are views on the nature of variability and the role of natural selection in its manifestation. Other

controversial issues such as the role of the direct action of the environment on evolution, the significance of parallelisms, the ageing of phylogenetic trends, the relation to vitalism etc. can be regarded as secondary. The most exact wording of the antagonism concerning the nature of variability was given by Sobolev: "Is it an accumulation and selection of random variations occurring in various directions, which in themselves do not predetermine a definite course of the phylogenetic process, which is governed by selection only, or is this elementary variability ordered and subject to certain laws, which predetermine, irrespectively of selection, the direction of evolution?" (Sobolev 1924, p. 24).

5. Orthogenetic concepts have acquired a solid reputation for vitalism. For this reason, they are often dismissed as hopelessly outdated. However, this is justified only in some cases (the works of Karl Ernst von Baer and Pierre Teilhard de Chardin). In general, the charges of vitalism directed at orthogenesis are unjust. There are traces of vitalism in almost any evolutionary concept. In particular, they can be seen the statements of modern Darwinists about the uniqueness of biology and the impossibility of reducing animate phenomena to those of inorganic world. The essence of vitalism is the idea that the living world has some unique features setting it aside from non-living matter. Most authors of the concepts of directed evolution tended towards an opposite viewpoint, trying to show that biological phenomena were determined by the same processes that operated in the inorganic world and to identify biological laws similar to physical, chemical and astronomical ones. Some versions of orthogenesis emphasised the similarity of organisms and machines, which is an idea alien to vitalism.
6. Concepts of directed evolution have appeared over the entire period of the existence of evolutionary biology, and keep emerging nowadays. At the same time, the main arguments have been repeated again and again. The following evidence has often been considered in favour of directed evolution: constraints on variation, purely age-related changes in the evolution of groups of organisms, preadaptations, parallelisms and convergences, maladaptive characters. The mechanisms of these phenomena are equally difficult to explain regardless of whether Darwinism or its alternatives are chosen as the theoretical framework. The supporters of the idea of directed evolution paid special attention to these phenomena and collected abundant material on them, in this way identifying promising research directions in evolutionary biology. In this regard, orthogenesis is a productive, though unrealised, trend rather than a dead end of the evolutionary thought.
7. Slow development of the idea of directed evolution is due to the following reasons. (1) Its supporters have been estranged, often trying to oppose their own work to that of others. (2) Their claims have often been poorly substantiated, loosing much in comparison with the circumstantial argumentation presented by Darwin. (3) The ideas of directed evolution have often been criticised with the use of philosophical, ideological and other non-scientific arguments, gaining in this way a heretical reputation. (4) The concepts of directed evolution are inconvenient because they focus on unresolved issues that traditional concepts tend to ignore.

References

Aguirre E (1958) Orthogenesis and the problem of biological evolution (La orthogenesis y el problema de la evolución biológica). Madrid Arbor 39(148):467–502

Akhnazarov EB (2002) The contours of evolution (Kontury evolyutsii). Nedra, St Petersburg

Baer KE von (1876) On the Darwinian teaching (Über Darwinsche Lehre). In: Studien aus dem Gebiete der Naturwissenschaften. Verlag der Caiferlichen Hofbuchhandlung H. Schmissdorf, St Petersburg, pp 235–479

Berg LS (1926) Nomogenesis or evolution determined by law, English edn. Constable & Co. ltd., London

Cope ED (1868) On the origin of genera. Proceedings of Academy of Natural Sciences of Philadelphia4, pp 242–305

Cope ED (1896) Primary factors of organic evolution. Open Court Publishing, Chicago

Croizat L (1958) Panbiogeography. 3 volumes. Published by the author, Caracas

Croizat L (1962) Spaceflight, Time, Form: the biological synthesis. Published by the author, Caracas

de Rosa D (1931) Ologenesis. A new theory of evolution and geographic distribution of living beings (L'Ologénèse. Nouvelle théorie de l'évolution et de la distribution géografique des êtres vivants). Libraire Félix Alcan, Paris

Eimer Th (1897) Orthogenesis of butterflies. An argument determines directed development and the impotence of natural selection for the origin of species. At the same time a rejoindre to August Weismann (Orthogenesis der Schmetterlinge. Ein Beweis bestimmt gerichteter Entwickelung und Ohnmacht der naturlichen Zuchtwahl bei der Artbildung. Zugleich eine Erwiderung an August Weismann). W. Engelmann, Leipzig

Gutmann WF (1994) Constructional constraints in the evolution: swimming tetrapods (Konstruktionszwänge in der evolution: schwimmende Vierfüsser). Nat Mus 124(6):165–189

Gutmann M, Weingarten M (1992) Principles of construction morphology and organismic theory of evolution (Grundlagen von Konstruktionsmorphologie und organismischerEvolutionstheorie). Aufsaetze und Reden senck naturf Ges 38:51–68

Gutmann M, Weingarten M (1994) Changes in the discussion of evolutionary theory: the abolishment of atomism in genetics (Veränderungen in der evolutiontheoretischen Diskussion: die Aufhebung des Atomismus in der Genetik). Nat Mus 124(6):189–196

Haacke W (1893) Form and inheritance. Developmental mechanichs of organisms (Gestaltung und Vererbung. Eine Entwickelungsmechanik der Organismen). T. O. Weigel Nachfolger, Leipzig

Kawamura K (2002) The origin of life from "the life of subjectivity". In: Fundamentals of life. Éditions scientifiques et médicales Elsevier SAS, pp 563–575

Kölliker RA (1864) On the Darwinian theory of creation (Über die Darwin'sche Schöpfungstheorie). Vortrag 1864 in Würzburg. Z Wiss Zool 14:174–186

Kölliker RA (1872) Morphology and developmental history of Pennatulid genera alongside general considerations concerning evolutionary theory (Morphologie und Entwickelungsgeschichte des Pennatulidenstammes nebst allgemeinen Betrachtungen zur Descendenzlehre). Christian Winter, Frankfurt a. M

Labbé A (1924) A new concept of adaptation: allelogenesis (Une conception nouvelle de l'adaptation: l'allélogénese). Revue scientifique 10:295–301

Lang WD (1916) Calcium carbonate and evolution in Polyzoa. Geological Magazine, Decade 6(3):73–77

Lang WD (1919) The Pelmatoporinae, an essay on the evolution of a group of cretaceous Polyzoa, Philosophical transactions of the Royal Society of London, Series B, vol 209, pp 191–228

Lima de Faria A (1988) Evolution without selection: form and function by autoevolution. Kindle Edition

Lima de Faria A (1995) Biological periodicity: its molecular mechanism and evolutionary implication. JAI Press, Greenwich (Conn)

Lwoff A (1943) Physiological evolution. A study of losses of function in microorganisms (L'évolution phisiologique. Étude des pertes de fonction chez les microorganisms). Actual Sc Industr 970

Nägeli C (1856) Individuality in nature with regard to plant world (Die Individualität in der Natur mit Berücksichtigung des Pflanzenreichs). Akademishe Vortrage, II. Verlag von Meyer & Zeller, Zurich, pp 171–212

Nägeli C (1865) Origin and concept of species in natural history (Entstehung und Begriff der Naturhistorischen Art), Im Verlage der koenigl Akademia, München

Nägeli C (1884) Mechanistic-physiological theory of evolution (Mechanischphysiologische Theorie der Abstammungslehre). Druck und Verlag von R. Oldenbourg, Munich and Leipzig

Osborn HF (1934) Aristogenesis, the creative principle in the origin of species. Amer Nat 68:193–235

Przibram H (1929) Apogenetical theory of the evolution of organisms (Théorie apogénétique de l'évolution des organisms). Revue générale des Sciences pures et appliquées 40:293–299

Schindewolf OH (1936) Paleontology, evolution and genetics: critics and synthesis (Paläontologie, Entwicklungslehre und Genetik: Kritik und Synthese). Borntraeger, 1936

Schindewolf OH (1950) Basic Questions in Palaeontology: Geologic Time, Organic Evolution and biological Systematics. (Grundfragen der Paläontologie. Geologische Zeitmessung, Organische Stammesentwicklung, Biologische Systematik). Schweizerbart'sche Verlagsbuchhandlung, Stuttgart

Sobolev DN (1924) The fundamentals of historical biogenetics (Nachala istoricheskoy biogenetiki). Gosudarstvennoe izdatel'stvo Ukrainy, Simferopol

Teilhard de Chardin P (1950) Concerning a remarkable case of orthogenesis of a group: evolution of Chinese siphneides (Sur un cas remarquable d'orthogénèse du groupe: l'évolution des siphnéides de Chine). Paléontologie et transformisme. Albin Michel, Paris, pp 169–173

Teilhard de Chardin P (1961) The phenomenon of man, Harper Torchbooks, The Cloister Library, Harper & Row Publishers

Vandel A (1964) Biospeleology. The biology of cave animals (Biospéologie. La biologie des Animaux Cavernicoles). Gauthier-Villars Éditeur, Paris. English edition: Vandel A (1965) biospeleology. The biology of cavernicolous animals. Pergamon Press, Oxford

Index

A
Abiotic and biotic factors, 130, 156
Action of environment, 9, 19, 21, 25, 26, 33, 35, 46, 59, 85, 92, 160, 177, 181, 191, 202, 203
Adaptation, 1, 9, 11, 14, 16, 20–22, 31, 32, 35, 36, 45, 48, 58, 59, 63, 69, 70, 79, 80, 87, 88, 90, 91, 109, 111, 120–122, 139–147, 151, 153–156, 165, 170, 182, 191
Allelogenesis, 48, 196, 201
Anticipation, 151–157
Apogenesis, 48, 196, 201
Aristogenes, 42, 44, 45, 152, 153
Aristogenesis, 42–44, 58, 104, 196, 201
Autoevolution, 1, 93, 195, 197, 201

B
Bio-events, 130–132
Biological synthesis, 90, 196, 201
Botany, 34, 69, 75, 89, 126, 145, 196

C
Change, 1, 3, 8–11, 13–19, 21–27, 31, 34, 38, 39, 41, 42, 45–49, 56, 58, 59, 61, 62, 69, 70, 73, 75, 77–81, 85–88, 92, 96, 98, 106–108, 110, 111, 117, 118, 121, 122, 130, 131, 133, 136, 140, 141, 145, 147, 153, 164, 165, 170, 173, 177, 178, 181–183, 193, 202, 203
Constraints, 9, 23, 24, 77, 81, 97, 103, 110, 112, 120, 157, 160–162, 164–167, 169, 173, 203
Constraints on variation, 23, 77, 81, 97, 111, 157, 167, 203
Contours of evolution, 96, 197, 201
Convergences, 35, 36, 57, 58, 64, 65, 67, 80, 104, 106, 151, 166, 203
Criticism, 9, 12, 17, 21, 26, 34, 41, 57, 61, 75, 77, 79, 81, 91, 92, 102, 103, 106, 110, 159, 161, 165, 177, 187, 188, 190, 191, 195

D
Darwinism, 1–3, 10, 16, 17, 27, 31–35, 37–39, 42, 45, 53, 54, 57, 58, 61–63, 68, 70, 73–79, 81, 85–87, 89, 91, 92, 95, 98, 101, 103–107, 110, 120, 132, 142, 144, 151, 155, 159, 162–164, 182, 187–190, 193, 202, 203
Dissidents, 101

E
Evolution, 1, 7, 31, 53, 73, 85, 103, 117, 129, 139, 151, 161, 177, 187
Evolutionary synthesis, 54, 73–81, 85, 88, 90, 103, 105, 106, 110, 112, 119, 143, 160, 161, 188, 194
Extinction, 10, 35, 41, 54, 55, 69, 108, 109, 111, 118, 119, 123, 129–135, 140, 195

F
Frankfurt theory, 91, 92

I. Popov, *Orthogenesis versus Darwinism*,
https://doi.org/10.1007/978-3-319-95144-7

G

Genetics, 17, 27, 31–33, 35, 45, 60, 61, 63, 64, 70, 74, 75, 106, 108, 143, 179, 180, 188, 189

H

Heresies, 21, 39, 62, 91, 104, 140, 156, 161
Historical biogenetics, 39, 41, 196, 201
Hollow curve, 125, 126

I

Idealistic morphology, 57, 66, 77, 193
Inadaptive evolution, 54, 139, 141
Inertia, 18, 120–122, 166
Inheritance, 8, 17, 24, 33, 42, 45, 49, 58, 61, 98, 177–184
Irreversibility, 40, 78, 121, 122

L

Lamarckism, 7, 9, 10, 13, 15, 25, 27, 33, 45, 46, 59, 62, 63, 73, 76, 85, 98, 111, 177–179, 182–184, 193, 201, 202
Laws, 2, 8–15, 17, 18, 20, 22–25, 27, 31, 34, 38–41, 47, 54, 60–62, 65, 68, 85, 92, 93, 102–104, 111, 117–126, 132, 139, 140, 151, 154, 155, 157, 159, 166, 169, 190, 191, 193, 203

M

Modern Darwinism, 2, 45, 53, 54, 58, 70, 73–78, 81, 103, 132, 163, 182
Mystics, 1, 80, 187–198

N

19th century, 2, 3, 7–28, 33, 41, 45, 53, 59, 68, 77, 119, 189, 190, 202
Nomogenesis, 1, 33–37, 39–41, 56, 58, 61–64, 93, 96, 102–104, 187–189, 196, 197, 201
Non-adaptive characters, 16, 24, 69, 87, 104
"Non-ageing" animals, 171, 172

O

Ologenesis, 1, 46–48, 90, 196, 197, 201
Organicism, 1, 86–88, 196, 197, 201
Orthogenesis, 1–3, 7, 9, 10, 16, 18–21, 26, 27, 33, 39, 44, 46, 48, 49, 53–70, 73–81, 86–94, 96, 101, 117–126, 135, 140–142, 152, 155, 156, 164, 166, 177, 184, 187–193, 195–197, 201–203

P

Palaeontology, 3, 35, 39, 40, 42, 45, 53, 54, 56, 60, 75, 89, 107, 117, 120, 122, 152, 196
Panglossian paradigm, 143, 144, 164
Parallelisms, 14, 35, 62–68, 80, 89, 97, 104–106, 126, 151–157, 161, 203
Preadaptation, 31, 59, 88, 141, 142, 203
Purposeful evolution, 67, 81

R

Recent history, 85–98, 101
Rules, 13, 117, 119, 120, 125, 146, 156, 182, 202

S

Senescence, 13, 41, 56, 68, 86–88, 108, 111, 129–136, 144, 146, 170–172
Shutayzey/subjectivity, 96, 197
Snouters, 169, 170
Specialization, 13, 40, 41, 59, 69, 78, 79, 87, 108, 117–119, 132, 134, 135
Support of orthogenesis, 63, 70, 80, 90, 101–112, 166
Survival, 37, 38, 121, 183
Synthesis, 27, 54, 57, 68, 69, 73–79, 85, 88, 90, 103, 106, 110, 112, 119, 143, 160, 161, 166, 188, 193, 194, 196, 197, 201

T

Teleology, 1, 10, 81, 87, 141–143, 188, 191, 192
Typostrophism, 85, 86, 110, 196, 201

U

Unsolved problems of evolutionary biology, 76

V

Vagueness, 75, 104
Variation, 9, 11, 12, 14–16, 19, 21–27, 32, 35, 36, 39, 41, 49, 53–55, 57–60, 62, 68,

69, 77, 79, 81, 93, 97, 105–107, 110, 111, 117, 119, 120, 122, 123, 125, 135, 154, 157, 159, 161–165, 167, 169, 173, 179, 181–183, 203
Vitalism, 1, 55, 92, 98, 188–192, 197, 203

Z
Zoology, 8, 42, 63, 69, 145, 196

MIX
Papier aus verantwortungsvollen Quellen
Paper from responsible sources
FSC® C105338

If you have any concerns about our products,
you can contact us on
ProductSafety@springernature.com

In case Publisher is established outside the EU,
the EU authorized representative is:
Springer Nature Customer Service Center GmbH
Europaplatz 3, 69115 Heidelberg, Germany

Printed by Libri Plureos GmbH
in Hamburg, Germany